Self-Organized 3D Tissue Patterns

Fundamentals, Design, and Experiments

T0313371

Xiaolu Zhu

Zheng Wang

JENNY STANFORD
PUBLISHING

Published by

Jenny Stanford Publishing Pte. Ltd.
Level 34, Centennial Tower
3 Temasek Avenue
Singapore 039190

Email: editorial@jennystanford.com
Web: www.jennystanford.com

British Library Cataloguing-in-Publication Data
A catalogue record for this book is available from the British Library.

Self-Organized 3D Tissue Patterns: Fundamentals, Design, and Experiments

Copyright © 2022 Jenny Stanford Publishing Pte. Ltd.

ISBN 978-981-4877-77-0 (Hardcover)
ISBN 978-1-003-18039-5 (eBook)

Contents

Preface

Tissue engineering applies principles and methods from engineering and life sciences to create artificial constructs to direct tissue regeneration or enhance tissues and organs. Structured scaffolds are widely useful for providing structures supporting cells to form 3D tissue. However, it is non-trivial to develop a scheme, which can robustly guide cells to self-organize into a tissue with desired 3D spatial structures. The self-organization of cells is a natural process that occurs in various biological bodies. In the field of engineering, we may expect that the process of self-organization can be rationally predicted and controlled. Moreover, it will be better that the self-organization of a huge number of cells can be governed by a mathematical framework.

This book first introduces the advances in tissue and organ regenerating using diverse technologies from different disciplines (Chapter 1). Then, it focuses on the multicellular or tissue structure formation via self-assembly of cells in 3D hydrogels (Chapter 2 and 3). Based on the hydrogel fabrication technique, the tailored inner interfaces inside hydrogels have been proposed to stimulate the tubular microtissue formation via a self-assembly scheme (Chapter 4), and the corresponding mathematical framework and the simulation model are also presented and discussed (Chapter 5). Furthermore, in order to develop a more elaborate and sophisticated regulating method for tuning the collective cellular behaviors, we also propose a hydrogel system with 3D distribution of clustered compositions, which can tune the multicellular elongations and aggregations (Chapter 6). This book offers the fundamentals and innovative designs for the 3D hydrogel system and discusses the representatively experimental results on the self-organized 3D Tissue patterns.

The authors are grateful for the helpful discussion with Prof. Ting-Hsuan Chen at City University of Hong Kong, Prof. Chih-Ming Ho at the University of California, Los Angeles (UCLA)，Prof. Tatiana Segura at Duke University and Prof. Alan Garfinkel and Prof. Yin Tintut at UCLA, while conducting the related projects. We hope this book will be useful for the readers in the interdisciplinary areas of engineering, biology, and life sciences.

Xiaolu Zhu
Zheng Wang
Hohai University, China

Chapter 1

Introduction

1.1 Research in Tissue Engineering

Tissue engineering applies principles and methods from engineering and life sciences to create artificial constructs to direct tissue regeneration or enhance tissues and organs [1, 2]. In the context of incessant development of tools and improvements, such as synthetic biology and genomic editing techniques [3], tissue engineering is playing a leading role as a multidisciplinary research branch. In order to achieve effective strategies to regenerate the functions of living tissues and organs, many cell-based tissue engineering methods have been proposed to heal or reconstitute and restore tissue functions for translational medicine. Tissue engineering emerged as a scientific field with great potentialities in the 1980s, and extensive work on it has been developed since then, which aims at regenerating skin, cartilage, bone, and many other prototypes of tissue and organ substitutes, such as nerve conduits, blood vessels, liver, and even heart. Early stage of tissue engineering was directly motivated by practical therapy demands in clinics, especially in the areas of skin replacement and cartilage repair. Skin grafts are the first engineered tissue constructs, and autologous skin grafts were the golden choice [4], yet the available donor sites of graft materials are limited. Similar

Self-Organized 3D Tissue Patterns: Fundamentals, Design, and Experiments
Xiaolu Zhu and Zheng Wang
Copyright © 2022 Jenny Stanford Publishing Pte. Ltd.
ISBN 978-981-4877-77-0 (Hardcover), 978-1-003-18039-5 (eBook)
www.jennystanford.com

to skin grafts, osteochondral transplantation techniques repair cartilage by placing the autologous donor tissue harvested from non-weight-bearing regions of the joint [5]. This technique is also limited by the rather short supply of donor tissue available. A following improved method is autologous chondrocyte implantation (ACI) that implants the autologous cells significantly expanded in vitro into the debrided cartilage defect. This method provides enough cells via in vitro expansion which alleviates the short supply of autologous donor tissue. However, sometimes there will be chondrocyte leakage without the retention by an artificial scaffold, providing an uneven chondrocyte distribution [6, 7]. Therefore, the structured scaffolds are required for better efficacy in tissue/organ regeneration.

Traditional structured scaffolds, sometimes called macro-scaffolds including exogenous scaffolds and decellularized 3D matrix scaffolds, may be the most widely investigated topic. Exogenous scaffolds, such as the synthetic porous scaffolds, provide spatial three-dimensional (3D) structure for seeded cells to adhere, spread, proliferate, differentiate, meanwhile to secrete extracellular matrix (ECM), and deliver biochemical factors across a 3D space [8–10], which could mimic the physicochemical, biological, and mechanical properties of native tissues/organs [11]. After cell seeding, the geometry and stiffness of the scaffold may influence the response of cells [12–16]. To reconstitute tissue architectural features, several studies have been devoted to fabricate various scaffolds with specific structures to guide cell spreading and growing [17–19]. Nevertheless, the structural complexity of the scaffolds is limited by the manufacture precision. On the other hand, decellularized 3D matrices obtained from solid cadaveric organ can retain the intact structure of natural tissues or organs. It advances the progress in building complex organs with vascular networks. In this technique, the original cells from the solid cadaveric organ could be decellularized using biochemical technology, and the obtained acellular natural scaffold with the original organ microarchitecture is then re-cellularized. The re-cellularized organ, including vascular network [20] seeded with autologous cells, is then transplanted into the patient. The tissue engineering based on decellularized natural scaffolds with great promise, yet still encounters the lack of donor organs with specific size and available functions for the patients. Although porcine organs are probably the potential substitutes for human organs because they are similar to human organs in size and function [21], currently the problems of immunological incompatibilities still exist after the important progress that the porcine endogenous retrovirus (PERV) in pigs was inactivated by using CRISPR-Cas9 [22].

In addition to structured scaffolds and decellularized matrices, a spectrum of approaches based on the self-organization process of cells have been studied and developed over the past three decades for tissue engineering and regenerative medicine. Central to this translational endeavor is taking advantage of cells' natural capacity to synthesize the ECM, self-assembly into tissue architectures and concurrently respond to signals [23–26]. Self-organization based engineering approaches relieve the limitations that the scaffold-based tissue engineering methods usually have probably included the non-synchronization of the scaffold degradation to neotissue (soft-tissue) formation and the concern of the immunogenicity due to scaffold creation, seeding, or degradation [27, 28]. Also, self-organization paradigm for fabricating tissues is usually realized by using the synthetic hydrogels as ECM. The hydrogels are easier to be produced relative to the decellularized 3D matrices that require donors. To get the controllable self-organization of cells, specifically synthesized ECM materials, external forces, specially tailored microfabrication [29], and the mathematical modeling for molecular and cellular activities [30], programmable control of endogenous gene networks [31] have been utilized as effective approaches.

The objective of this chapter is to describe advances in tissue and organ morphogenesis using diverse technologies from different disciplines. Traditional tissue grafting and cell implantation for skin or cartilage at early development stage of tissue engineering is described first. Then the commonly used structured scaffolds with gross morphology or structure features of native tissues are reviewed. The synthetic scaffolds and decellularized 3D matrices are the two main categories of the structured scaffolds here, and their features, developments, and limitations are, respectively, discussed. Subsequently, the massive efforts pursuing the controllable self-organizing scheme for tissue and organ regeneration are reviewed. The advantages and performances of the guidance by computational approaches for tissue/organ morphogenesis driven by self-organization scheme are also discussed in the following chapters. Rationally regulating self-organization exhibits high potential for complex tissue and organoid regeneration, which is an important step toward clinical translation.

1.2 Traditional Tissue Grafting and Typical Cell Implantation for Skin or Cartilage

Tissue engineering that combines knowledge from molecular biology, materials science, biomechanics, and medicine, intends to produce tissue constructs to repair or replace native tissues compromised by trauma, pathology, or age [32]. The brief history and progress of tissue engineering can be found in Berthiaume et al. [33]. Early studies were motivated by practical therapy demands, especially in the areas of skin replacement and cartilage repair, which are the most representative tissue engineering treatments. Skin grafts are the first engineered tissue constructs, and the early-stage skin grafts [27, 34–36] were mainly used for wounds having a diameter larger than 1 cm or that extend deep into the dermis, thereby requiring special treatment for closure. Autologous skin grafts remain the best choice [4], but the available donor sites are limited. Alternatively, the early tissue-engineered skin constructs were also produced by in vitro coculturing the keratinocytes isolated from patients [34–36] with a feeder layer of mouse mesenchymal cells. The keratinocytes grew ex vivo and expanded the coverage area a thousand-fold in several weeks. This achievement led to the first cell-based tissue-engineered product called Epicel, for treating patients suffering from catastrophic burn injuries. But this product does not have a dermis and has a thickness filled with a few cells, which limited its common usage. To make a skin-equivalent tissue with full thickness, researchers developed a composite skin product named Apligraf that reconstituted both dermis and epidermis [27]. Since then, other analogous products incorporating bovine type I collagen sponge as substrates have also been developed. However, because living cells are used, immunological rejection may occur. Some alternative types of skin grafts were developed, which contained no extra living cells before they were placed at the wound surface. Their main mission was to guide and stimulate the body's repair and regenerative processes. They used a porous acellular matrix located at the wound surface, thus allowing the migration and revascularization of host cells such as endothelial cells, neural cells, and fibroblasts into it [37].

Similar to skin grafts, the development of cartilage grafts is also driven by clinical demands. The demand for engineered and regenerative cartilages has been growing because of the increasing prevalence of degenerative joint diseases (e.g., osteoarthritis). Additionally, the spontaneous repair for the injured articular cartilage is still limited even for young and healthy individuals [38]. Articular cartilage is a thick, avascular, aneural tissue that located at the ends of long bones [39, 40]. Since the articular cartilage naturally possesses no vasculature, it is hard to regrow and get the neotissue (soft-tissue) regenerated. Artificial cartilage repairing approaches have been investigated for over 30 years, and important successes have been achieved in clinics. The transplant of autologous donor tissue harvested from a non-weight-bearing region to the damaged site can help to repair cartilages. However, similar to the case with autologous skin grafts, this approach is subject to the availability of cells and donor sites. At early stage, microfracture [41] was introduced in the late 1980s and early 1990s to treat the cartilage defects. This technique can enhances migration of mesenchymal stem cells (MSCs) from bone marrow to the site of a cartilage defect by penetrating the subchondral bone [42]. However, microfracture often leads to fibrocartilage formation and treatment failure was expected in a term more than 5 years after surgery [43, 44]. A following improved method is ACI. In this method, a chondrocyte population, collected from the patient, is expanded in vitro, yielding ~12–48 million cells; then the chondrocyte population is implanted into the debrided cartilage defect. This technique showed prolonged improvement compared to autologous osteochondral transplantation in excess of 10 years postoperatively [45, 46]. For the allograft transplantation, there was a significant loss of viability after 15 years [47].

The strategy of tissue grafting and cell implantation could work for skin and cartilage because these tissues have a relative simpler inner microstructure and function mechanism. They do not require extensive vascularization and other significant tissue processes. Directly performing the tissue grafting from the autologous donor site is still facing the short supply of autologous donor tissue. Moreover, the skin substitutes also may lack native skin appendages and innate cell types, including sweat glands, sebaceous glands, and melanocytes, as well as langerhans or dendritic cells [4, 48]. The cell implantation-based method could alleviate the short supply of autologous donor tissue, because the cells could be significantly expanded in vitro before implantation. Yet, the cell implantation site would require the suitable clinical operation that still needs to be more standardized, and the patient-specific and cartilage defect-specific factors should be comprehensively considered. Sometimes, there will be chondrocyte leakage without the retention by an artificial scaffold, providing a uneven chondrocyte distribution, and it consequently results in graft hypertrophy [6, 7]. Nevertheless, large-scale cohort studies are needed to be further investigated to develop rational methodology, aiming at advancing the consistency and long-term effectiveness of cartilage repairs.

1.3 Synthetic Structured Scaffold and Decellularized 3D Matrices

The lack of a supportive scaffold material to guide cellular organization and matrix synthesis may account for the limited regeneration of inner tissue structures and slow migration and revascularization of host cells as well as the leakage of the implanted cells [7, 49]. In recent years, reliable therapies have been sought to regenerate damaged tissues and organs [50–53] based on structured scaffolds. Increasing complex and sophisticated scaffolds are widely used as engineered ECM. To recapitulate architectural features of native tissues in microenvironments, some attempts have been made to fabricate the scaffolds with specific structure to guide cellular spreading [9], assembling layers of cultured cell sheets [54, 55], or electrically patterning cells by using a concentric-stellate-tip electrode array as a template [56]. To engineer structure and function of living tissues and organs, researchers designed and synthesized diverse types of scaffolds over the past few decades. The structured scaffolds are widely used in biological tissue engineering, and the manufacturing of scaffolds is gradually toward the direction of more complex and fine features [8, 9, 57–60], including the photopolymer scaffold structure fabricated by excimer laser based on stereolithography production [59], the porous metal scaffold structure fabricated by laser assisted perforated method [18, 61], metal scaffold made via solid free-forming process [18, 62], as well as polymer material scaffolds manufactured using rapid prototyping technology [28]. In the reconstruction of the characteristics of biological tissue structure in the microenvironment, various supporting scaffolds promoted the cellular adhesions through their unique structural characteristics and surface coating processing. The scaffold prompted cells to grow in an arrangement in line with the geometric features of scaffolds, and the structured scaffolds may also degrade complying with the design requirement. These endeavors made an important contribution to the development of tissue engineering.

Although the synthetic structured scaffolds or sometimes called porous scaffolds mentioned above can serve as candidates for tissue regeneration, the supporting scaffolds do not fully mimic the native physiological microenvironment, and sometimes involve immunogenic reaction. Considering the skin regeneration, even some porous acellular matrices/templates were utilized to promote the cellular proliferation and tissue regeneration, the skin substitutes were slow revascularized and sometimes poorly attached to the wound bed [33]. Therefore, the structured scaffolds-based tissue engineering has its own limitations. In clinical application, the degradation macro-scaffolds or called porous scaffolds are usually not synchronized to neo-tissue formation, remodeling, and integration into functional properties [24]. Furthermore, regardless how sophisticatedly the scaffolds are designed and fabricated, they can cause eliciting immunogenic reaction or other unforeseen complications in the course of scaffold seeding and degradation [28]. The position of cells will change during the growth process; the geometrical characteristics of resulting multicellular architecture sometimes spatially deviate from the original design.

As previously noted, some acellular matrix template made from the decellularized donor tissue could keep the native matrix components yet contained no cells, which prevented an allogeneic immunological response. From this perspective, the decellularized 3D matrices have special advantages and could also serve as a good candidate for tissue or organ scaffold. The decellularization method was utilized very early on investigating cell functions, phenotypes and behaviors in structured matrices with natural ECM components, and the application targets of the method included skin [63], adipose tissue [64], mammary gland [65], and more recently, the bones [66, 67]. The decellularization does not only help build the simply structured tissue, but also can advance the progress in building complex organs with vascular networks. As we know, the organs with relatively large size such as the heart, lung, liver, and kidney require immediate access to adequate blood supply to obtain nutrients and oxygen after transplantation. Thus, the intact vascular network is an essential part of the transplanted organs, which maintains cell survival and organ function. Therefore, in order to obtain a supporting scaffold closer to natural ones, one suitable solution is direct use of a natural cadaveric organ to build the native and entire ECM. After the original cells from the solid cadaveric organ are decellularized using detergents, enzymes, physical agents, or other chemicals, the acellular natural scaffold with the original organ microarchitecture, including vascular network and naturally-derived ECM, is obtained [20], and then the acellular natural scaffold could be re-cellularized, followed by the transplantation of it into the body [68]. After the perfusion-based decellularization, by using recellularization technique to seed the scaffold with autologous cells, whole organ engineering has been performed in the heart [69, 70], lung [71], kidney [72], and liver [68, 73], and has been tested in pre-clinical in vivo studies. Although the decellularization/recellularization technique holds great promise, challenges remain because there is still a shortage of candidate donor organs with specific size and available functions for the patients, and it is difficult for preserving the entire/native ECM components without any residual cytoplasmic and nuclear materials after decellularization [74]. More detailed summary of diverse scaffolds and fabrications linked to tissue and organ development can be found in [77, 78].

1.4 Tissue and Organoid Morphogenesis by Regulated Self-Organization Process

As the statements above, although an increased interest has been in the use of synthetic structured scaffold and decellularized matrices that retain the natural microarchitecture as the scaffold, those acellular structured scaffolds are produced through conventional manufacturing methods in mechanical or chemical engineering, and it is thus difficult to integrate them with living cells simultaneously during the manufacturing or synthetic processes because UV exposure, laser melting, electrospinning with high voltages and/or solvent infusion may be harmful to cells. The previously mentioned biodegradability and toxicity of the synthetic structured scaffold after transplantation may also compromise the progress in clinical applications. Although the decellularized natural scaffolds have great promise, there is still a paucity of candidate donor organs with specific size and available functions for the patients, and the optimized protocols for decellularization/recellularization need to be organ-specifically and patient-specifically established, giving rise to high costs and clinic experience requirements [62].

To go beyond the limitation, attention has been tuned to the use of self-organizing properties and the innate regenerative capability of the tissue/organism formation [23, 24]. Rational control of the self-organization process in tissue engineering could relieve the limitations that porous scaffolds or decellularized natural scaffolds usually have, probably including the non-synchronization of the scaffold degradation to neotissue formation and the concern of the immunogenicity due to scaffold creation, seeding, or degradation [27, 28]. What's more, the cell migration, proliferation, and differentiation may disorganize and frustrate the pre-design of porous scaffolds or decellularized natural scaffolds. Currently, the controllable methodology for creating desired 3D self-organized multicellular structures in vitro is on the way, and it is important for promoting the development of tissue engineering tissue engineering and regenerative medicine [75, 76]. The following chapters will describe the advances in self-organization of cells and tissues in a more rational and controllable manner, and the intrinsic laws for cellular self-organization and regulation methods on multicellular structure and tissue morphogenesis will be elucidated, analyzed, and discussed.

References

1. Ladoux, B.; Mège, R.-M. (2017). Mechanobiology of collective cell behaviours, *Nature Reviews Molecular Cell Biology,* **18**, pp. 743.

2. Tiwari, S.; Patil, R.; Bahadur, P. (2018). Polysaccharide based scaffolds for soft tissue engineering applications, *Polymers,* **11**, pp. 1.

3. Singh, V.; Braddick, D.; Dhar, P. K. (2017). Exploring the potential of genome editing CRISPR-Cas9 technology, *Gene,* **599**, pp. 1–18.

4. MacNeil, S. (2007). Progress and opportunities for tissue-engineered skin, *Nature,* **445**, pp. 874.

5. Lee, C.; Grodzinsky, A.; Hsu, H.; Martin, S.; Spector, M. (2000). Effects of harvest and selected cartilage repair procedures on the physical and biochemical properties of articular cartilage in the canine knee, *J. Orth. Res.,* **18**, pp. 790–799.

6. Caplan, N.; Kader, D. F. (2000). Two- to 9-Year outcome after autologous chondrocyte transplantation of the knee, *Clin. Orthop. Relat. Res.,* **1**, pp. 212–234.

7. Gooding, C. R.; Bartlett, W.; Bentley, G.; Skinner, J. A.; Carrington, R.; Flanagan, A. (2006). A prospective, randomised study comparing two techniques of autologous chondrocyte implantation for osteochondral defects in the knee: Periosteum covered versus type I/III collagen covered, *Knee,* **13**, pp. 203–210.

8. Hutmacher, D. W.; Sittinger, M.; Risbud, M. V. (2004). Scaffold-based tissue engineering: Rationale for computer-aided design and solid free-form fabrication systems, *Trends Biotechnol,* **22**, pp. 354–62.

9. Tsang, V. L.; Bhatia, S. N. (2004). Three-dimensional tissue fabrication, *Adv Drug Deliv Rev,* **56**, pp. 1635–47.

10. Hutmacher, D. W. (2001). Scaffold design and fabrication technologies for engineering tissues-state of the art and future perspectives, *J Biomater Sci Polym Ed.,* **12**, pp. 107–124.

11. Di Luca, A.; Van Blitterswijk, C.; Moroni, L. (2015). The osteochondral interface as a gradient tissue: From development to the fabrication of gradient scaffolds for regenerative medicine, *Birth Defects Res C Embryo Today,* **105**, pp. 34–52.

12. Levy-Mishali, M.; Zoldan, J.; Levenberg, S. (2009). Effect of scaffold stiffness on myoblast differentiation, *Tissue Engineering Part A,* **15**, pp. 935–944.

13. Discher, D.; Janmey, P.; Wang, Y.-I. (2005). Tissue cells feel and respond to the stiffness of their substrate, *Science,* **310**, pp. 1139 –1143.

14. Maleki, H.; Shahbazi, M.-A.; Montes, S.; Hosseini, S. H.; Eskandari, M. R.; Zaunschirm, S.; Verwanger, T.; Mathur, S.; Milow, B.; Krammer, B.; Huesing, N. (2019). Mechanically strong silica-silk fibroin bioaerogel: A hybrid scaffold with ordered honeycomb micromorphology and multiscale porosity for bone regeneration, *ACS Appl. Mater. Interfaces,* **11**, pp. 17256–17269.

15. Ameer, J. M.; Pr, A. K.; Kasoju, N. (2019). Strategies to tune electrospun scaffold porosity for effective cell response in tissue engineering, *Journal of Functional Biomaterials,* **10**(3), 30:1–21.

16. Leong, M. F.; Rasheed, M. Z.; Lim, T. C.; Chian, K. S. (2010). In vitro cell infiltration and in vivo cell infiltration and vascularization in a fibrous, highly porous poly(D,L-lactide) scaffold fabricated by cryogenic electrospinning technique, *Journal of Biomedical Materials Research Part A,* **91A**, pp. 231–240.

17. Ma, H.; Feng, C.; Chang, J.; Wu, C. (2018). 3D-printed bioceramic scaffolds: From bone tissue engineering to tumor therapy, *Acta Biomater,* **79**, pp. 37–59.

18. Yusop, A. H.; Bakir, A. A.; Shaharom, N. A.; Abdul Kadir, M. R.; Hermawan, H. (2012). Porous biodegradable metals for hard tissue scaffolds: A review, *Int J Biomater,* **2012**, pp. 641430.

19. Youssef, A.; Hollister, S. J.; Dalton, P. D. (2017). Additive manufacturing of polymer melts for implantable medical devices and scaffolds, *Biofabrication,* **9**, pp. 012002.

20. Arenasherrera, J. E.; Ko, I. K.; Atala, A.; Yoo, J. J. (2013). Decellularization for whole organ bioengineering, *Biomedical Materials,* **8**, pp. 014106.

21. Deschamps, J. Y.; Roux, F. P.; Gouin, E. (2010). History of xenotransplantation, *Xenotransplantation,* **12**, pp. 91–109.

22. Niu, D.; Wei, H.-J.; Lin, L.; George, H.; Wang, T.; Lee, I.-H.; Zhao, H.-Y.; Wang, Y.; Kan, Y.; Shrock, E.; Lesha, E.; Wang, G.; Luo, Y.; Qing, Y.; Jiao, D.; Zhao, H.; Zhou, X.; Wang, S.; Wei, H.; Güell, M.; Church, G. M.; Yang, L. (2017). Inactivation of porcine endogenous retrovirus in pigs using CRISPR-Cas9, *Science,* **357**, pp. 1303–1307.

23. Mironov, V.; Trusk, T.; Kasyanov, V.; Little, S.; Swaja, R.; Markwald, R. (2009). Biofabrication: A 21st century manufacturing paradigm, *Biofabrication,* **1**, 022001: 1–16.

24. Lei, Y.; Gojgini, S.; Lam, J.; Segura, T. (2011). The spreading, migration and proliferation of mouse mesenchymal stem cells cultured inside hyaluronic acid hydrogels, *Biomaterials,* **32**, pp. 39–47.

25. Bedzhov, I.; Zernickagoetz, M. (2014). Self-organizing properties of mouse pluripotent cells initiate morphogenesis upon implantation, *Cell,* **156**, pp. 1032–44.

26. Athanasiou, K. A.; Eswaramoorthy, R.; Hadidi, P.; Hu, J. C. (2013). Self-organization and the self-assembling process in tissue engineering, *Annu. Rev. Biomed. Eng.,* **15**, pp. 115–136.

27. Bell, E.; Ehrlich, H. P.; Buttle, D. J.; Nakatsuji, T. (1981). Living tissue formed in vitro and accepted as skin-equivalent tissue of full thickness, *Science,* **211**, pp. 1052–1054.

28. Jakab, K.; Norotte, C.; Marga, F.; Murphy, K.; Vunjak-Novakovic, G.; Forgacs, G. (2010). Tissue engineering by self-assembly and bio-printing of living cells, *Biofabrication,* **2**, pp. 022001.

29. Zhu, X.; Gojgini, S.; Chen, T. H.; Teng, F.; Fei, P.; Dong, S.; Segura, T.; Ho, C.-M. (2016). Three-dimensional tubular structure self-assembled by vascular mesenchymal cells at stiffness interfaces of hydrogels, *Biomedicine & Pharmacotherapy,* **83**, pp. 1203–1211.

30. Zhu, X.; Gojgini, S.; Chen, T. H.; Fei, P.; Dong, S.; Ho, C.-M.; Segura, T. (2017). Directing three-dimensional multicellular morphogenesis by self-organization of vascular mesenchymal cells in hyaluronic acid hydrogels, *Journal of Biological Engineering,* **11**, pp. 12.

31. Teague, B. P.; Guye, P.; Weiss, R. (2016). Synthetic morphogenesis, *Cold Spring Harb Perspect Biol,* **8**, pp. 1–15.

32. Athanasiou, K. A. (2013). Self-organization and the self-assembling process in tissue engineering, *Annu. Rev. Biomed. Eng.,* 2013, **15**(1), pp. 115–136.

33. Berthiaume, F.; Maguire, T. J.; Yarmush, M. L. (2011). Tissue engineering and regenerative medicine: History, progress, and challenges, *Annu Rev Chem Biomol Eng,* **2**, pp. 403–430.

34. Green, H.; Kehinde, O.; Thomas, J. (1979). Growth of cultured human epidermal cells into multiple epithelia suitable for grafting, *Proc. Natl. Acad. Sci. USA,* **76**, pp. 5665–5668.

35. O'Connor, N.; Mulliken, J.; Banks-Schlegel, S.; Kehinde, O.; Green, H. (1981). Grafting of burns with cultured epithelium prepared from autologous epidermal cells, *Lancet,* **317**, pp. 75–78.

36. Rheinwald, J.; Green, H. (1975). Serial cultivation of strains of human epidermal keratinocytes: The formation of keratinizing colonies from single cells, *Cell,* **6**, pp. 331–43.

37. Yannas, I. V.; Burke, J. F.; Orgill, D. P.; Skrabut, E. M. (1982). Wound tissue can utilize a polymeric template to synthesize a functional extension of skin, *Science,* **215**, pp. 174–176.

38. Buckwalter, J. A.; Mankin, H. J. (1998). Articular cartilage: Degeneration and osteoarthritis, repair, regeneration, and transplantation, *Instr Course Lect,* **47**, pp. 487–504.

39. Stockwell, R. A. (1971). The interrelationship of cell density and cartilage thickness in mammalian articular cartilage, *J. Anat.,* **109**, pp. 411–421.

40. Hunziker, E. B.; Quinn, T. M.; Häuselmann, H. J. (2002). Quantitative structural organization of normal adult human articular cartilage, *Osteoarthritis & Cartilage,* **10**, pp. 564–572.

41. Dae Kyung, B.; Kyoung Ho, Y.; Jun, S. S. (2006). Cartilage healing after microfracture in osteoarthritic knees, *Arthroscopy the Journal of Arthroscopic & Related Surgery,* **22**, pp. 367–374.

42. Kreuz, P. C.; Steinwachs, M. R.; Erggelet, C.; Krause, S. J.; Konrad, G.; Uhl, M.; Südkamp, N. (2006). Results after microfracture of full-thickness chondral defects in different compartments in the knee, *Osteoarthritis & Cartilage,* **14**, pp. 1119–1125.

43. Goyal, D.; Keyhani, S.; Lee, E. H.; Hui, J. H. P. (2013). Evidence-based status of microfracture technique: A systematic review of level I and II studies, *Arthroscopy-the Journal of Arthroscopic & Related Surgery,* **29**, pp. 1579–1588.

44. Gudas, R.; Gudaitė, A.; Mickevičius, T.; Masiulis, N.; Simonaitytė, R.; Čekanauskas, E.; Skurvydas, A. (2013). Comparison of osteochondral autologous transplantation, microfracture, or debridement techniques in articular cartilage lesions associated with anterior cruciate ligament injury: A prospective study with a 3-year follow-up, *Arthroscopy-the Journal of Arthroscopic & Related Surgery,* **29**, pp. 89–97.

45. Bentley, G.; Biant, L. C.; Vijayan, S.; Macmull, S.; Skinner, J. A.; Carrington, R. W. (2012). Minimum ten-year results of a prospective randomised study of autologous chondrocyte implantation versus mosaicplasty for symptomatic articular cartilage lesions of the knee, *Journal of Bone & Joint Surgery British Volume,* **94**, pp. 504.

46. Lars, P.; Vasiliadis, H. S.; Mats, B.; Anders, L. (2010). Autologous chondrocyte implantation: A long-term follow-up, *Am. J. Sports Med.,* **38**, pp. 1117.

47. Gross, A. E.; Shasha, N.; Aubin, P. (2005). Long-Term followup of the use of fresh osteochondral allografts for posttraumatic knee defects, *Clin. Orthop. Relat. Res.,* **435**, pp. 79–87.

48. Metcalfe, A. D.; Ferguson, M. W. (2007). Tissue engineering of replacement skin: The crossroads of biomaterials, wound healing, embryonic development, stem cells, and regeneration, *Journal of the Royal Society Interface,* **4**, pp. 413.

49. Makris, E. A.; Gomoll, A. H.; Malizos, K. N.; Hu, J. C.; Athanasiou, K. A. (2015). Repair and tissue engineering techniques for articular cartilage, *Nature Reviews Rheumatology,* **11**, pp. 21–34.

50. Mimeault, M.; Batra, S. K. (2006). Concise review: Recent advances on the significance of stem cells in tissue regeneration and cancer therapies, *Stem Cells,* **24**, pp. 2319–45.

51. Eiraku, M.; Takata, N.; Ishibashi, H.; Kawada, M.; Sakakura, E.; Okuda, S.; Sekiguchi, K.; Adachi, T.; Sasai, Y. (2011). Self-organizing optic-cup morphogenesis in three-dimensional culture, *Nature,* **472**, pp. 51–6.

52. Rizzoti, K.; Lovell-Badge, R. (2011). Regenerative medicine: Organ recital in a dish, *Nature,* **480**, pp. 44–6.

53. Bashur, C. A.; Venkataraman, L.; Ramamurthi, A. (2012). Tissue engineering and regenerative strategies to replicate biocomplexity of vascular elastic matrix assembly, *Tissue Engineering Part B: Reviews,* **18**, pp. 203–217.

54. Shimizu, T.; Yamato, M.; Kikuchi, A.; Okano, T. (2003). Cell sheet engineering for myocardial tissue reconstruction, *Biomaterials,* **24**, pp. 2309–16.

55. Yang, J.; Yamato, M.; Shimizu, T.; Sekine, H.; Ohashi, K.; Kanzaki, M.; Ohki, T.; Nishida, K.; Okano, T. (2007). Reconstruction of functional tissues with cell sheet engineering, *Biomaterials,* **28**, pp. 5033–43.

56. Ho, C.-T.; Lin, R.-Z.; Chang, W.-Y.; Chang, H.-Y.; Liu, C.-H. (2006). Rapid heterogeneous liver-cell on-chip patterning via the enhanced field-induced dielectrophoresis trap, *Lab on a Chip,* **6**, pp. 724–734.

57. Li, X.; Feng, Y. F.; Wang, C. T.; Li, G. C.; Lei, W.; Zhang, Z. Y.; Wang, L. (2012). Evaluation of biological properties of electron beam melted Ti6Al4V implant with biomimetic coating in vitro and in vivo, *PLoS ONE,* **7**, pp. e52049.

58. Bhumiratana, S.; Vunjak-Novakovic, G. (2012). Concise review: Personalized human bone grafts for reconstructing head and face, *Stem Cells Transl Med,* **1**, pp. 64–9.

59. Beke, S.; Anjum, F.; Tsushima, H.; Ceseracciu, L.; Chieregatti, E.; Diaspro, A.; Athanassiou, A.; Brandi, F. (2012). Towards excimer-laser-based stereolithography: A rapid process to fabricate rigid biodegradable photopolymer scaffolds, *Journal of the Royal Society, Interface,* **9**, pp. 3017–26.

60. Weijie, Z.; Qin, L.; Dichen, L.; Kunzheng, W.; Zhongmin, J.; Weiguo, B.; Yaxiong, L.; Jiankang, H.; Ling, W. (2014). Cartilage repair and subchondral bone reconstruction based on three-dimensional printing technique, *Chinese Journal of Reparative and Reconstructive Surgery,* **28**, pp. 318–324.

61. Tan, L.; Gong, M.; Zheng, F.; Zhang, B.; Yang, K. (2009). Study on compression behavior of porous magnesium used as bone tissue engineering scaffolds, *Biomed Mater, 4*, pp. 015016.

62. Nguyen, T. L.; Staiger, M. P.; Dias, G. J.; Woodfield, T. B. F. (2011). A novel manufacturing route for fabrication of topologically-ordered porous magnesium scaffolds, *Adv. Eng. Mater., 13*, pp. 872–881.

63. Pushpoth, S.; Tambe, K.; Sandramouli, S. (2009). The use of alloderm in the reconstruction of full-thickness eyelid defects, *Orbit 27*, pp. 337–340.

64. Flynn, L. E. (2010). The use of decellularized adipose tissue to provide an inductive microenvironment for the adipogenic differentiation of human adipose-derived stem cells, *Biomaterials, 31*, pp. 4715–4724.

65. Wicha, M. S.; Lowrie, G.; Kohn, E.; Bagavandoss, P.; Mahn, T. (1982). Extracellular matrix promotes mammary epithelial growth and differentiation in vitro, *Proc. Natl. Acad. Sci. USA, 79*, pp. 3213–3217.

66. Darja, M.; Iván Marcos, C.; Sarindr, B.; Ana, K.; Petros, P.; Geping, Z.; Spitalnik, P. F.; Grayson, W. L.; Gordana, V. N. (2012). Engineering bone tissue from human embryonic stem cells, *Proc Natl Acad Sci USA, 109*, pp. 8705–8709.

67. Giuseppe Maria, D. P.; Iván, M. C.; David John, K.; Dana, A.; Linshan, S.; Gordana, V. N.; Darja, M. (2013). Engineering bone tissue substitutes from human induced pluripotent stem cells, *Pnas, 110*, pp. 8680–8685.

68. Meng, F.; Almohanna, F.; Altuhami, A.; Assiri, A. M.; Broering, D. (2019). Vasculature reconstruction of decellularized liver scaffolds via gelatin-based re-endothelialization, *Journal of Biomedical Materials Research Part A, 107*, pp. 392–402.

69. Meşină, M.; Mîndrilă, I.; Mesina, C.; Obleagă, C. V.; Istrătoaie, O. (2019). A perfusion decellularization heart model-an interesting tool for cell-matrix interaction studies, *Journal of Mind and Medical Sciences, 6*, pp. 137–142.

70. Wainwright, J. M.; Czajka, C. A.; Patel, U. B.; Freytes, D. O.; Tobita, K.; Gilbert, T. W.; Badylak, S. F. (2010). Preparation of cardiac extracellular matrix from an intact porcine heart, *Tissue Eng Part C Methods, 16*, pp. 525–532.

71. Tebyanian, H.; Karami, A.; Motavallian, E.; Samadikuchaksaraei, A.; Arjmand, B.; Nourani, M. (2019). Rat lung decellularization using chemical detergents for lung tissue engineering, *Biotechnic & Histochemistry, 94*, pp. 214–222.

72. Hussein, K. H.; Saleh, T.; Ahmed, E.; Kwak, H. H.; Park, K. M.; Yang, S. R.; Kang, B. J.; Choi, K. Y.; Kang, K. S.; Woo, H. M. (2018). Biocompatibility and hemocompatibility of efficiently decellularized whole porcine kidney for tissue engineering, *Journal of Biomedical Materials Research Part A,* **106**, pp. 2034–2047.

73. Baptista, P. M.; Siddiqui, M. M.; Genevieve, L.; Rodriguez, S. R.; Anthony, A.; Shay, S. (2011). The use of whole organ decellularization for the generation of a vascularized liver organoid, *Hepatology,* **53**, pp. 604–17.

74. Gilbert, T. W.; Freund, J. M.; Badylak, S. F. (2009). Quantification of DNA in biologic scaffold materials, *J. Surg. Res.,* **152**, pp. 135–139.

75. Griffin, D. R.; Archang, M. M.; Kuan, C.-H.; Weaver, W. M.; Weinstein, J. S.; Feng, A. C.; Ruccia, A.; Sideris, E.; Ragkousis, V.; Koh, J.; Plikus, M. V.; Di Carlo, D.; Segura, T.; Scumpia, P. O. (2021) Activating an adaptive immune response from a hydrogel scaffold imparts regenerative wound healing. *Nat. Mater.,* **20**, pp. 560–569.

76. Darling, N. J.; Xi, W.; Sideris, E.; Anderson, A. R.; Pong, C.; Carmichael, S. T.; Segura, T. (2020) Click by click microporous annealed particle (MAP) scaffolds. *Adv .Healthcare Mater.,* **9**, 1901391:1–10.

77. Zhu, X.; Wang, Z.; Teng, F. (2021). A review of regulated self-organizing approaches for tissue regeneration. *Prog. Biophys. Mol. Biol.* **167**, pp. 63–78.

78. Mota, C.; Camarero-Espinosa, S.; Baker, M. B.; Wieringa, P.; Moroni, L. (2020). Bioprinting: From tissue and organ development to in vitro models. *Chem. Rev.* **120**, pp. 11032–11092.

Chapter 2

Fundamentals of Three-Dimensional Cell Culture in Hydrogels

2.1 Introduction

For the therapies for regenerating damaged biological tissues or organs [1–4], many different cell-based treatments have been utilized to heal, reconstruct, or restore tissue function, with the intention of developing efficient tactics to regenerate the functions of living organs and tissues [2, 5–7]. Sophisticated scaffolds are widely used to function as artificial extracellular matrices, and various attempts have been made to produce scaffolds, which have specific structures to regulate cell spreading, package layers of cultured cell sheets, deposit cells directly, or translocate cells [8–10]. When the scaffolds are applied, nevertheless, their degradations are usually not synchronized with the new tissue formation at the expense of remodeling and integration, thus affecting the functional properties [11]. Further studies show that no matter how complicated the scaffolds were designed and made, they can lead to eliciting immunogenic reactions or other unpredicted complications in the process of scaffold seeding and degradation [12]. The scaffold geometry may also have an impact on the cellular differentiation through the mechanical cues of changing cellular figure [13] or scaffold stiffness [14]. Ultimately, it can be founded that the best way

Self-Organized 3D Tissue Patterns: Fundamentals, Design, and Experiments
Xiaolu Zhu and Zheng Wang
Copyright © 2022 Jenny Stanford Publishing Pte. Ltd.
ISBN 978-981-4877-77-0 (Hardcover), 978-1-003-18039-5 (eBook)
www.jennystanford.com

may rely on the inherent self-organizing properties of cells, tissues, and organisms themselves [11, 15], because in vitro self-organization of multicellular structures might appear similar to those that occur in the course of organ development. It will help to recapitulate morphological structure and organization in native organisms.

The self-organization can be divided into two types: two-dimensional (2D) and three-dimensional (3D) self-organization of cells. Both 2D and 3D cell self-organizations show their superiority in addressing the immunogenic reaction issues compared to macro structural scaffolds. 3D cellular self-organization, nevertheless, has its own obvious advantages compared to the 2D case. Some typical differences between 2D and 3D multicellular self-organization could be summarized as follows:

(i) Cells cultured in 3D environments can dramatically impact cellular organization, polarity, and drug responsiveness, compared to 2D case [16].

(ii) Commonly, cells grown in 2D monolayer cultures may lose the tissue-specific properties. The 3D cell culture models, however, will likely serve to link the cell-based assays and animal studies, aiming to reduce experimental uncertainties arising from 2D cultures [17].

(iii) In terms of structural and functional properties, multicellular spheroids in 3D resemble real tissues better than conventional monolayer cultures [17].

(iv) 2D and 3D cultured cells can show significantly different gene-expression profiles, suggesting that cellular attachment to extracellular matrix (ECM) influences cellular responses to mechanical stresses [18].

Therefore, controlling 3D cellular self-organization plays a great role for facilitating fundamental biomedical research and applied biomaterial development.

Here, we elucidate a 3D cell culture model for studying cellular self-assembly, in which vascular mesenchymal stem cells (MSCs) are encapsulated into 3D modified hyaluronic acid (HA) hydrogel. It is a typical 3D culture model to bring the basic information on the operating protocol and the obtained experimental results. Here, the MSC is a non-hematopoietic, multipotent progenitor cell, first identified in the marrow stroma, is well known to evolve in the human artery wall especially associating with atherosclerotic lesions.

Now they are harvested from a variety of sources of tissues, including adipose tissue, periosteum, liver, spleen, fetal blood, and diseased tissues, and they can differentiate into multiple lineages, including fat, cartilage, muscle, vascular tissue, and osteoblasts (bone cells) in vivo and in vitro [19–23]. MSCs, capable of differentiating in vitro and in vivo to mesenchymal lineages, have potentials for cell-based therapy because animal studies have shown that transplanted MSC stimulate angiogenesis in several different tissues [24]. MSCs have several properties such as being trophic, immunomodulatory, anti-inflammatory, anti-microbial, and anti-apoptotic [25]. Experiments in vitro show that vascular-derived MSCs can display self-organized calcified patterns. Activator/inhibitor dynamics under Turing-style mechanism contributed to these patterns, with matrix GLA protein acting as an activator, and bone morphogenetic protein 2 (BMP2) acting as an inhibitor [26].

3D cell culture model serves as a suitable fundamental platform for studying the cellular self-organization that mimics the natural development of tissues or trauma repairs. Cellular self-organization is triggered inside the 3D culture model integrated with molecules and cells, and sustainingly driven by a reaction-diffusion process of morphogens, which could be proteins or cytokines. In this chapter, we demonstrated the self-organized multicellular 3D structure and accommodation of the hydrogel components can be achieved in 3D culture regime. Based on the 3D HA hydrogels culture model, more features possessed by the self-organized cellular structure that have not been exhibited in 2D dishes, are comparatively elucidated.

2.2 Experimental Methods for 3D Culture of Cells

The experimental methods involved in 3D cell culture in hydrogels have some common protocols, but also vary when different hydrogels or cells involved. Here, we used the acrylated hyaluronic acid (HA-AC) hydrogel as our typical ECM. This hydrogel has several components and the proportion of each component can be tuned by researchers based on the specific consideration of their unique experiment design. The amount of the formed hydrogel does not require a large volume, but only 10–30 μL is needed in many practical cases. The readers may be suggested to make each hydrogel sample of low volume and fabricate more hydrogel samples. The adequate number of hydrogel samples will be used to complete more tests at once. For choosing the culture plates, the plate with small wells (such as 96–well plate) would be suggested because small wells not only provide a compact space for the small hydrogel samples, but also make the cytokines or growth factors secreted by cells, reaching the "working concentration" as soon as possible. These experiences can be validated just by comparing the cell growth situations in two wells added with different volumes of culture medium, such as 100 µL and 1 mL, if other conditions are all the same. The detailed experimental methods are described below. The descriptions include typical methods and materials used in the experiments for modifying the HA hydrogel and 3D culture in hydrogels.

2.2.1 Cell Culture

Vascular mesenchymal cells (VMCs), a type of MSCs, were isolated and cultured as described in literatures [27–29]. The cells were grown in Dulbecco's Modified Eagle's Medium (DMEM, Invitrogen) supplemented with 15% heat-inactivated fetal bovine serum (FBS) and 1% penicillin/streptomycin (10,000 I.U./10,000 µg/mL; all from Mediatech, VA). The cells were incubated at 37 °C in a humidified incubator (5% CO_2 and 95% air), and dissociated from the dish bottom using 0.25% trypsin-EDTA following standard protocols and passaged every three days.

2.2.2 Hyaluronic Acid Modification

Acrylated hyaluronic acid (HA-AC) was prepared by using a two-step synthesis. HA (1.0 g, 60 kDa) was reacted with 18.0 g of adipic dihydrazide (ADH) at a pH of 4.75 in the presence of 2.0 g of 1-ethyl-3-[3-dimethylaminopropyl] carbodiimide hydrochloride (EDC) overnight, and purified through dialysis in DI water for a week. The purified intermediate (HA-ADH) was lyophilized and stored at −20 °C until used. Around 16% of the carboxyl groups were modified with ADH based on the trinitrobenzene sulfonic acid (TNBSA, Pierce, Rockford, Illinois) assay.

2.2.3 VMCs-Laden HA Hydrogel Formation

HA hydrogels were formed by Michael addition of bis-cysteine containing MMP-degradable crosslinker onto HA-AC pre-functionalized with cell adhesion peptides (RGD). The final gel was swelled in culture media before being placed inside 96-well plates for the following long-term culture. The mechanical properties of the hydrogels can be controlled by changing the HA concentration or the crosslinking density, "r", which is defined as moles of -SH from the crosslinkers over moles of -ACs from the HA-ACs. The schematic diagram and four detailed drawings (A, B, C, and D) are shown in Figure 2.1.

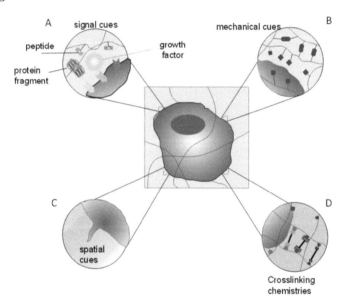

Figure 2.1 HA hydrogel can be fabricated with polymers, adhesive peptides, and crosslinking peptides, which incorporates multi-cues to regulate encapsulated cell behavior. (A) Bioactive signals are typically incorporated into HA hydrogels to guide cell behavior. Some signals, like peptides and antibody fragments, can be covalent binding with HA backbone. (B) The mechanical properties could be tuned via utilizing degradable crosslinkers, which can control cell mechanosensing. (C) Spatial cues, such as porosity control, patterned bioactive signals, and topographical patterns, contributed to regulating cell migration. (D) Crosslinking chemistries serve as new tools to crosslink the hydrogel network covalently [30].

2.2.4 Fixation and Fluorescent Staining

Before fixing the cells inside the hydrogel, the gel samples were washed twice in the micro-well plate with prewarmed phosphate-buffered saline (PBS). Then the cells in gel were fixed by 4% paraformaldehyde (PFA) solution for 40 min at room temperature. The samples were stained by fluorescent dyes that can render the F-actins (Alexa Fluor 488 phalloidin, Invitrogen) and the nuclei (DAPI, Invitrogen) inside the cells. For F-actin staining, 1% bovine serum albumin (BSA) was added to the staining solution of Alexa Fluor 488 phalloidin to reduce nonspecific background staining, and then the staining solution was pipetted to cover the cell-laden hydrogel sample in microwells for 25 min at room temperature. For nuclei staining, the solution of 300 nM DAPI in PBS was added to microwells containing the cell-laden hydrogel samples and the sample was incubated for 5 min.

2.2.5 3D Visualization Using Selective Plane Illumination Microscopy

After encapsulating in the hydrogel samples, cells were stained by the phalloidin and DAPI. The HA-gel samples were immersed into a low gelling-temperature agarose solution (37.5°C) which was composed of 0.5% (w/w) agarose and the rest of PBS. The agarose was gelled at room temperature and made the HA-gel sample mounted inside it. Then, the cuvette was placed onto the automatic stage of the selective plane illumination microscopy (SPIM) platform. The sample was optically sliced when the stage was programmed to travel at 1 μm or 2 μm per step. The SPIM imaging platform is more suitable for observing samples in large vision field using 4X or 10X objectives, which usually has higher imaging quality than common confocal microscopy. The obtained image sequences were stacked and processed by commercial 3D reconstruction software (Amira 5.2 Resolve RT, trial version). The basic principle of SPIM is stated in the reference [30] in Chapter 1.

2.2.6 Measuring Cell Proliferation in 3D HA Hydrogels

The Alamar blue (Invitrogen, USA) assay was used to quantify the proliferation of cell lines inside the 3D culture medium. This assay is based on detection of metabolic activity via an oxidation-reduction indicator, which can fluoresces. Besides the absorbance and fluorescence intensity is in proportion to the number of active cells. 20 mL of Alamar blue dye was mixed with 100 mL phenol red free DMEM and added to each gel-containing well and incubated at 37°C for 4 h. The proliferation rate multiplied in number in the first three days of culture.

2.3 Results and Discussion for 3D Microtissue Patterns Emerged in HA Hydrogels

The cell populations can grow in the modified HA hydrogel, and the cells can sprout, spread, proliferate, and connect with other cells in 3D hydrogel matrices.

2.3.1 Generation of 3D Structures Composed of Aggregated Cells

As shown in Figure 2.2, 2D cells are self-organized into multicellular patterns with special shapes that may resemble tissue architectures. At the beginning, the cells were roughly and evenly distributed on 2D surface and they were in a free state with random distribution (Figure 2.2a). To study cell patterns, VMCs were traced by green fluorescent protein (Figure 2.2b). Ten days later, cells aggregated into curved ridges. The cells in ridges also combined with surrounding cells and formed connections. The non-uniformity of cell density was obvious, and the aligned cellular orientations indicated the tendency of cell migration. Meanwhile, filament-like connection links indicated the start of self-organization (Figure 2.2c). After 15 days, the multicellular patterns with inhomogeneous cell density were more obvious and cell migration was almost reaching the steady state. Patterns were formed via the self-organization of multicellular structures, and the enlargement of local pattern represented a dense artificial tissue microstructure.

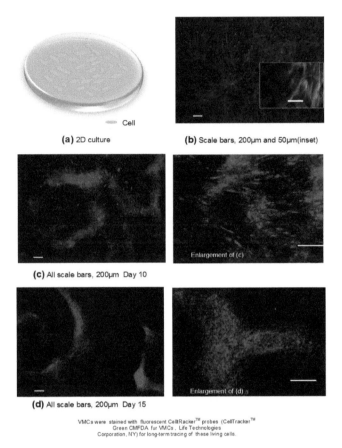

(a) 2D culture

(b) Scale bars, 200μm and 50μm(inset)

(c) All scale bars, 200μm Day 10

Enlargement of (c)

(d) All scale bars, 200μm Day 15

Enlargement of (d)

VMCs were stained with fluorescent CellRacker™ probes (CellTracker™ Green CMFDA fur VMCs , Life Technologies Corporation, NY) for long-term tracing of these living cells.

Figure 2.2 Cells cultured in 2D dishes. (a) A sketch of culture environment shows a monolayer of randomly oriented VMCs of nearly uniform density. (b) Green fluorescent protein was used to track VMCs. (c) The initial formation of local alignment. (d) Ridges of high cell concentration (dark red areas). WMCs were stained with fluorescent CellTracker probes. (CellTracker Green CMFDA, Life Technologies Corporation, NY, for long-term tracing of living cells).

We demonstrated that VMCs have good growth in 3D HA hydrogels. The 3D morphologies, composed of multicellular structures crowded with aggregated cells, were formed during cell migration and proliferation, as shown in Figure 2.3. Bossing-like multicellular tissues closely linked to the surrounding multicellular structures. Multicellular aggregation with cells branching out looks like a stellate structure. Inevitably, holes exist between the connection links. Strong branches of bossing-like tissues present a long rod shape, while thin branches look like filament. At first, the cells were roughly evenly distributed in the 3D HA hydrogels, with little contact between each other. Inside the 3D hydrogel, cells spread, proliferated, and migrated, with the increased culture time. The cells could degrade ECM and gradually opened up the routes so as to migrate and then self-organized into multicellular structures. Eventually, cells bound to each other and formed spatial connections. The structure composed of cell population represents a solid reticulation construct, which can serve as a unit block for resembling natural tissue structures.

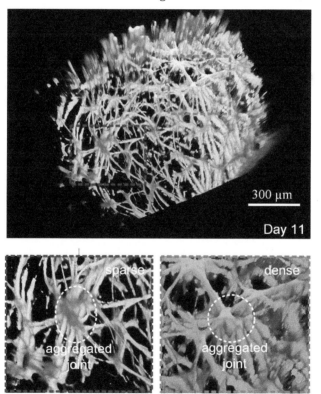

Figure 2.3 3D morphology of micro-tissue consisting of multicellular network in HA hydrogels. SPIM technique was used to map the distribution of fluorescent intensity in 3D, which was reconstructed from volumetric data from cross-sectional views (sliced images). Two local regions have been enlarged as shown in the below figures, which depict the aggregated joints. The two enlarged figures show the sparse network and the dense network at different regions.

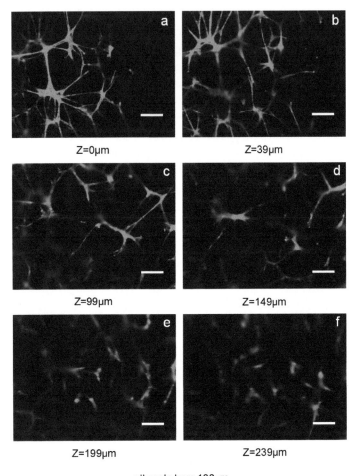

Z=0μm

Z=39μm

Z=99μm

Z=149μm

Z=199μm

Z=239μm

all scale bars 100μm.

Figure 2.4 VMCs in 3D HA hydrogels in vitro. 500 confocal images were taken to show the cross-section views of each local morphologies of aggregated cells from $z = 0$ μm to $z = 499$ μm at the interval of 1μm. Due to the limitation of light depth, images get vaguer with the increase of z coordinates. Six images were chosen to show the vivid cross-section views of cultured 3D multicellular structures.

Sectional views of HA hydrogel sample show that the detail structure with tissue-specific properties proves that it is 3D construct obviously (Figure 2.4). Filament-like linking structures are connected with single cells and multicellular constructs. Connections between cells became diversified including straight or curved geometries and crisscrossed in 3D matrix. According to the scale bars, we roughly conjectured the size of visible multicellular structures, connection links, and holes.

2.3.2 Influence of Component Proportion on Self-Organization of VMCs in HA Hydrogel

HA hydrogels with 3.0% and 3.5% HA and either 100, 400, 500, or 1000 µM RGD were used. RGD peptides, adhesion points in the gels, are ligands for integrin receptors on cells. RGD concentration also could affect the cell proliferation to some extent. From Figure 2.5, low RGD concentration resulted in enhanced proliferation especially in 3.0% HA at RGD concentration = 100 µM (left). The proliferation of the living cells in 3.0% and 3.5% HA hydrogels with 100, 400, 500, and 1000 µM RGD were all observed during the culture, yet cells proliferated relatively slow at higher HA concentrations.

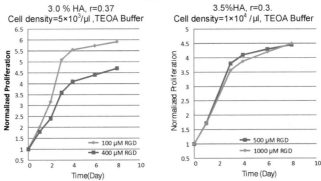

Figure 2.5 Test example for the proliferation of cells in 3.0% HA at two different RGD concentrations (100 µM, 400 µM) with cell density = 5000/µL, and $r = 0.37$ (left). 3.5% HA hydrogels at two different RGD concentrations (500 µM, 1000 µM) with $r = 0.3$, and cell density = 1×10^4/µL (right).

Some other testing experiments also performed. The VMCs initially presented a uniform distribution. The cells can spread, migrate, and proliferate inside the modified HA hydrogel. A few cells had small sprouting projections. After 2–3 days, the VMCs apparently spread out with finger-like projections and usually kept this morphology in the following days (Figure 2.6). VMCs started remarkable interaction with neighboring cells as early as day 2, which became more pronounced between days 4 and 5 (Figure 2.6). 3D multicellular networks formed consisted of a number of branches and sub-branches. Gradually, many cells aggregated and connected with each other and finally formed a large multicellular dense structure. This dense network structure had an irregular and rough appearance in the 3D matrix. As shown in Figure 2.6c, the proliferation became more significant when the lower concentration of RGD was employed, and the amount of proliferated cells could nearly reach five times the initial cell amount at day 3. Changing crosslinking density (the *r* value) can also regulate the proliferation of cells, and a lower crosslinking density leads to higher proliferation of cells. This can be attributed to the relative larger pores inside the inner structure of hydrogel, which provides the cells more local spaces (microcosmic surface area and volume) and less astricts during cellular spread, elongation, and proliferation.

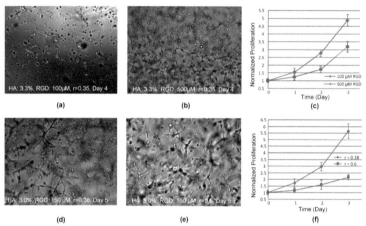

Figure 2.6 Variation of morphology of the VMC culture under different experimental conditions. (a,b) Cell morphology on day 4 at different RGD concentrations (3.3% HA, *r* = 0.35). (c) Proliferation of cells over time at different RGD concentrations (3.3% HA, *r* = 0.35). (d,e) Cell morphology on day 5 at different crosslinker concentrations that characterized by the parameter *r* (3.0% HA, 150 μM RGD). (f) Proliferation of cells over time at different *r* values (3.0 % HA, 150 μM RGD).

2.3.3 Cytotoxicity of Modified Dextran Hydrogels and Cellular Proliferation Measurement in Hydrogels

When the substrate materials in cell culture are transferred from 2D substrates to 3D hydrogel matrices, the viability and proliferation of cells can be varied, because the dimension of the cell-substrate interaction is changed, the style for cells to uptake the nutrient and self-organization will be also changed [31]. Therefore, the cytotoxicity and cellular proliferation measurement should be conducted again in 3D culture after the measurement has been done in 2D culture. The cellular viability reflects the chance for cells to survive and the biological compatibility of the hydrogel material. The proliferation of cells reflects the ability of cells to continuously grow and the chance for forming the larger scale multicellular structure that needs more and more cells.

An increased uptake rate of substance, including nutrients or fluorescent probes, due to the larger cell surface exposed to the surrounding environment, could facilitate more effective diffusion through the cellular membrane. It may result in a reduced effective concentration of the agent or assay used in 3D matrices compared to 2D cultures, which then results in a dilution factor that could account for the difference in the IC50 calculated [32, 33] and/or increased cell resistance to toxic/chemotherapeutic agents.

The cellular growth performance is usually influenced by the cell–matrix interactions, especially in the study on the stem cells behaviors in 3D matrices. Like most stem cells, the immortalized C2C12 cells are capable of self-renewing to produce the same cells [34]. Under specific conditions, C2C12 cells can differentiate into several types of cells, such as adipocytes, muscle cells, and osteoblasts. Hence, C2C12 cells have pluripotency and their behaviors could be representative for a part of stem cells, especially myogenic stem cells. In addition, the fate of the C2C12 cells is usually easier to be influenced by the extracellular microenvironment, especially the mechanical property of hydrogels. Additionally, NIH-3T3 cells, derived from mouse embryonic fibroblasts, tend to fill spaces within tissues and form ECM [35], whereas endothelial cells form into tubular architectures with a central lumen that is essential to blood vessel function. Fibroblasts also have been frequently used as recipients of DNA to detect cell-transforming genes for analyzing the tumorigenic potential of various patterns of gene expression. Therefore, the C2C12 and NIH-3T3 cells are used as typical cell models to demonstrate the cytotoxicity of the hydrogels and the cellular proliferations in the 3D hydrogels.

A live/dead test has been conducted on 3T3 and C2C12 on day 0, day 3, day 6, and day 9, respectively with initial cell density of 5000/μL. The results are shown in Figure 2.7. Green spots represent living cells, and red spots represent dead cells. The results showed that NIH–3T3 fibroblast and C2C12 cells in the 2D petri dish showed higher survival rates that those in the 3D dextran hydrogel with homogenous distributions of RGD peptides (Figure 2.7b). The number of green spots was larger than the number of red spots in the image (Figure 2.7a). It indicated that both 3T3 and C2C12 cells can keep viability in 3D dextran hydrogel for days. With the extension of culturing time, the 'green' cells appeared to aggregate, and grew into specific structures in the hydrogel (Figure 2.7a). This evolution indicated that cells in such hydrogel material can keep capacities of proliferation and grow into multicellular structures. The mean and standard deviations of the data were conducted (Figure 2.7b), including the cell survival rates of 3T3 and C2C12 cells, respectively, on day 0, day 3, day 6, and day 9. Both 3T3 and C2C12 cells kept a survival rate of over 75%, which increased over the first three days and were kept stable during the entire test.

The proliferation of 3T3 and C2C12 cells was studied by manual counting on day 3, day 6, and day 9, respectively. The initial cell quantity of each sample was 150,000. The results showed that the proliferation of 3T3 and C2C12 cells in 3D dextran hydrogel increased, but was hardly doubled during the cultivation for nine days (Figure 2.7c). Although 3T3 and C2C12 cells in the 3D hydrogel in a 96-well plate exhibited lower proliferation rate than the cells on 2D surface in a six-well plate with the same initial cell quantity (Figure 2.8), the results also informed us that such hydrogel material can hold 150,000 cells in 3D matrices in a microwell with a bottom area of 32 mm^2 initially, and keep cells proliferating continuously. During cultivation in hydrogel, the proliferation of 3T3 and C2C12 cells showed different regularities. The proliferation of 3T3 fibroblasts increased continuously for nine days (Figure 2.7c), while the proliferation of C2C12 was increased continuously and higher than that of 3T3 over the first six days, but it decreased obviously on day 9 (Figure 2.7c).

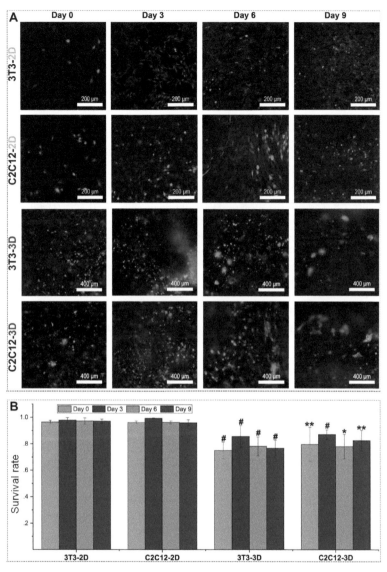

Figure 2.7 Viability and proliferation of cells in 2D and 3D. (A) Live/dead test conducted on 3T3 and C2C12 in a 2D petri dish (TCPS control) and 3D dextran hydrogel. (B) Survival rates of 3T3 and C2C12 from day 0 to day 9. The data were presented by mean ± SD. The data were presented by mean ± SD; *$p < 0.05$, **$p < 0.03$, and #$p < 0.01$ versus the corresponding 2D control samples.

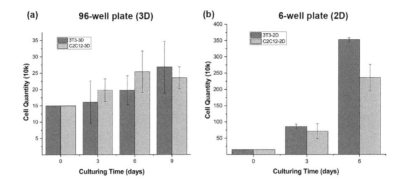

Figure 2.8 The proliferation of 3T3 and C2C12 cells in 3D and 2D with an initial cell amount of 150,000. (a) Proliferation of 3T3 and C2C12 cells in 3D homogenous dextran hydrogels in the 96-well plate. (b) Proliferation of 3T3 and C2C12 cells on the 2D bottom surface of the six-well plate. The data was presented as mean ± SD.

2.3.4 Discussion on Self-Organization of Cells in 3D Hydrogel with Quantitatively Tunable Components

3D culture of MSCs plays important roles in trauma repairs. Usually, inflammation at the trauma site can impede stem cells from generating native tissues or driving organ repair in many cases of musculoskeletal trauma [25]. MSCs have the ability of trauma repair and anti-inflammatory, and can secret growth factors and anti-inflammatory molecules. Besides, proteins like IGF-1, IL-6 expressed by MSCs are recognized to enhance pro-survival Akt (protein kinase B) expression, which are supposed to enhance cell viability [36, 37]. The 3D culture of MSCs is a fundamental platform for regenerative medicine or disease treatment.

Further, we proposed that the observed morphologies of 3D multicellular architectures are associated with a Turing-type mechanism. The mechanism has been used to explain the pattern formation in nature (such as patterns on the skins of tigers), but it has not been proved to be capable of interpreting and guiding the 3D multicellular architecture formation in vitro. As first proposed by Turing [38], biological patterns emerging from instability of a homogeneous steady state in biology can often be modeled mathematically by postulating "morphogens" that react chemically and also diffuse. The evolution of complex multicellular patterns can be interpreted by the reaction-diffusion framework, in which chemicals produced by cells interact as activators and inhibitors (morphogen pairs) and diffuse through the matrices at distinct rates. Reaction-diffusion equations and modeling morphogen pairs have been used to analyze pattern formation in many chemical and biological systems [39–43]. The existence of morphogens and their effects on tissues in vertebrates have been proved [44]. Specific morphogens for creating biological patterns have been identified, including transforming growth factor (TGF) β family members, proteins of the Wnt and Hedgehog families, and retinoic acid [40–46]. Certain genes modulate the molecular signaling of morphogens [40], which presents a more complicated working mechanism of reaction and diffusion of morphogens. More explanation and expatiation will be conducted in following chapters.

2.4 Summary

This chapter elucidates a typical 3D cell culture model for studying cellular self-assembly, in which vascular MSCs are encapsulated into 3D modified HA hydrogel or the muscle stem cells (C2C12) are embedded into the modified dextran hydrogel. It provides the basic information on the operating scheme and protocol of 3D cell culture in hydrogels. The obtained 3D micro-tissue patterns are influenced by the matrix property and component proportion in hydrogels, which lay a foundation for understanding following chapters.

Acknowledgments

This work is supported by the National Natural Science Foundation of China (Grant Nos. 51875170 and 51505127) and Fundamental Research Funds for the Central Universities (Grant Nos. B200202225, 2018B22414, and 2015B04414) of China.

References

1. Mimeault, M.; Batra, S. K. (2006). Concise review: Recent advances on the significance of stem cells in tissue regeneration and cancer therapies, *Stem Cells,* **24**, pp. 2319–45.

2. Eiraku, M.; Takata, N.; Ishibashi, H.; Kawada, M.; Sakakura, E.; Okuda, S.; Sekiguchi, K.; Adachi, T.; Sasai, Y. (2011). Self-organizing optic-cup morphogenesis in three-dimensional culture, *Nature,* **472**, pp. 51–6.

3. Rizzoti, K.; Lovell-Badge, R. (2011). Regenerative medicine: Organ recital in a dish, *Nature,* **480**, pp. 44–6.

4. Bashur, C. A.; Venkataraman, L.; Ramamurthi, A. (2012). Tissue engineering and regenerative strategies to replicate biocomplexity of vascular elastic matrix assembly, *Tissue Engineering Part B: Reviews,* **18**, pp. 203–217.

5. Huh, D.; Matthews, B. D.; Mammoto, A.; Montoya-Zavala, M.; Hsin, H. Y.; Ingber, D. E. (2010). Reconstituting organ-level lung functions on a chip, *Science,* **328**, pp. 1662–8.

6. Berthiaume, F.; Maguire, T. J.; Yarmush, M. L. (2011). Tissue engineering and regenerative medicine: History, progress, and challenges, *Annu Rev Chem Biomol Eng,* **2**, pp. 403–430.

7. Haycock, J. W. (2011). 3D cell culture: A review of current approaches and techniques, *Methods Mol Biol,* **695**, pp. 1–15.

8. Odde, D. J.; Renn, M. J. (2000) Laser-guided direct writing of living cells. *Biotechnol Bioeng,* **67,** pp. 312–8.

9. Mironov, V.; Boland, T.; Trusk, T.; Forgacs, G.; Markwald, R. R. (2003). Organ printing: Computer-aided jet-based 3D tissue engineering, *Trends Biotechnol,* **21**, pp. 157–61.

10. Ho, C.-T.; Lin, R.-Z.; Chang, W.-Y.; Chang, H.-Y.; Liu, C.-H. (2006). Rapid heterogeneous liver-cell on-chip patterning via the enhanced field-induced dielectrophoresis trap, *Lab on a Chip,* **6**, pp. 724–734.

11. Lei, Y.; Gojgini, S.; Lam, J.; Segura, T. (2011). The spreading, migration and proliferation of mouse mesenchymal stem cells cultured inside hyaluronic acid hydrogels, *Biomaterials,* **32**, pp. 39–47.

12. Jakab, K.; Norotte, C.; Marga, F.; Murphy, K.; Vunjak-Novakovic, G.; Forgacs, G. (2010). Tissue engineering by self-assembly and bio-printing of living cells, *Biofabrication,* **2**, pp. 022001.

13. Kilian, K. A.; Bugarija, B.; Lahn, B. T.; Mrksich, M. (2010). Geometric cues for directing the differentiation of mesenchymal stem cells, *Proc. Natl. Acad. Sci. USA,* **107**, pp. 4872–4877.

14. Levy-Mishali, M.; Zoldan, J.; Levenberg, S. (2009). Effect of scaffold stiffness on myoblast differentiation, *Tissue Engineering Part A* **15**, pp. 935–944.

15. Mironov, V.; Trusk, T.; Kasyanov, V.; Little, S.; Swaja, R.; Markwald, R. (2009). Biofabrication: A 21st century manufacturing paradigm, *Biofabrication,* **1**, 022001: 1–16.

16. Danino, T.; Volfson, D.; Bhatia, S. N.; Tsimring, L.; Hasty, J. (2011). In-silico patterning of vascular mesenchymal cells in three dimensions, *PLoS ONE,* **6**, pp. e20182.

17. McMahon, J. M.; Wells, K. E.; Bamfo, J. E.; Cartwright, M. A.; Wells, D. J. (1998). Inflammatory responses following direct injection of plasmid DNA into skeletal muscle, *Gene Ther,* **5**, pp. 1283–1290.

18. Treier, M.; Gleiberman, A. S.; O'Connell, S. M.; Szeto, D. P.; McMahon, J. A.; McMahon, A. P.; Rosenfeld, M. G. (1998). Multistep signaling requirements for pituitary organogenesis in vivo, *Genes Dev.,* **12**, pp. 1691–1704.

19. Chung, B. G.; Choo, J. (2010). Microfluidic gradient platforms for controlling cellular behavior, *Electrophoresis,* **31**, pp. 3014–3027.

20. Brunet, L. J.; McMahon, J. A.; McMahon, A. P.; Harland, R. M. (1998). Noggin, cartilage morphogenesis, and joint formation in the mammalian skeleton, *Science,* **280**, pp. 1455–1457.

21. McMahon, J. A.; Takada, S.; Zimmerman, L. B.; Fan, C. M.; Harland, R. M.; McMahon, A. P. (1998). Noggin-mediated antagonism of BMP signaling is required for growth and patterning of the neural tube and somite, *Genes Dev.,* **12**, pp. 1438–1452.

22. Anker, P. S.; Scherjon, S. A.; Kleijburg-van der Keur, C.; Noort, W. A.; Claas, F. H.; Willemze, R.; Fibbe, W. E.; Kanhai, H. H. (2003). Amniotic fluid as a novel source of mesenchymal stem cells for therapeutic transplantation, *Blood,* **102**, pp. 1548–9.

23. Itsykson, P.; Ilouz, N.; Turetsky, T.; Goldstein, R. S.; Pera, M. F.; Fishbein, I.; Segal, M.; Reubinoff, B. E. (2005). Derivation of neural precursors from human embryonic stem cells in the presence of noggin, *Mol. Cell. Neurosci.,* **30**, pp. 24–36.

24. Yang, R.; Song, Y. (2015). Bifurcation analysis of a diffusive activator–inhibitor model in vascular mesenchymal cells, *International Journal of Bifurcation and Chaos,* **25**, pp. 1530026.

25. Murphy, M. B.; Moncivais, K.; Caplan, A. I. (2013). Mesenchymal stem cells: Environmentally responsive therapeutics for regenerative medicine, *Exp Mol Med.,* **45**, pp. e54.

26. Yochelis, A.; Tintut, Y.; Demer, L. L.; Garfinkel, A. (2008). The formation of labyrinths, spots and stripe patterns in a biochemical approach to cardiovascular calcification. *N. J. Phys.,* **10**, pp. 78–106.

27. Bostrom, K.; Watson, K. E.; Horn, S.; Wortham, C.; Herman, I. M.; Demer, L. L. (1993). Bone morphogenetic protein expression in human atherosclerotic lesions, *J. Clin. Invest.,* **91**, pp. 1800–1809.

28. Hsiai, T. K.; Cho, S. K.; Reddy, S.; Hama, S.; Navab, M.; Demer, L. L.; Honda, H. M.; Ho, C. M. (2001). Pulsatile flow regulates monocyte adhesion to oxidized lipid-induced endothelial cells, *Arterioscler. Thromb. Vasc. Biol.,* **21**, pp. 1770–1776.

29. Wong, P. K.; Yu, F. Q.; Shahangian, A.; Cheng, G. H.; Sun, R.; Ho, C. M. (2008). Closed-loop control of cellular functions using combinatory drugs guided by a stochastic search algorithm, *Proc. Natl. Acad. Sci. USA,* **105**, pp. 5105–5110.

30. Lam, J.; Truong, N. F.; Segura, T. (2014). Design of cell–matrix interactions in hyaluronic acid hydrogel scaffolds, *Acta Biomater.,* **10**, pp. 1571–1580.

31. Bonnier, F.; Keating, M. E.; Wrobel, T. P.; Majzner, K.; Baranska, M.; Garcia-Munoz, A.; Blanco, A.; Byrne, H. J. (2015). Cell viability assessment using the Alamar blue assay: a comparison of 2D and 3D cell culture models, *Toxicol. In Vitro,* **29**, pp. 124–31.

32. Millerot-Serrurot, E.; Guilbert, M.; Fourré, N.; Witkowski, W.; Said, G.; Van Gulick, L.; Terryn, C.; Zahm, J.-M.; Garnotel, R.; Jeannesson, P. (2010). 3D collagen type I matrix inhibits the antimigratory effect of doxorubicin, *Cancer Cell International,* **10**, pp. 26.

33. Godugu, C.; Singh, M. (2016). AlgiMatrix™-based 3D cell culture system as an in vitro tumor model: an important tool in cancer research. In: Strano, S. (ed) *Cancer Chemoprevention.* Methods in Molecular Biology, vol. 1379, pp. 117–128. Humana Press, New York, NY. https://doi.org/10.1007/978-1-4939-3191-0_11.

34. Yaffe, D.; Saxel, O. (1977). Serial passaging and differentiation of myogenic cells isolated from dystrophic mouse muscle, *Nature,* **270**, pp. 725–727.

35. Salem, A. K.; Stevens, R.; Pearson, R. G.; Davies, M. C.; Tendler, S. J. B.; Roberts, C. J.; Williams, P. M.; Shakesheff, K. M. (2002). Interactions of 3T3 fibroblasts and endothelial cells with defined pore features, *J. Biomed. Mater. Res.,* **61**, pp. 212–217.

36. Mitsiades, C.; Mitsiades, N.; Poulaki, V.; Schlossman, R.; Akiyama, M.; Chauhan, D.; Hideshima, T.; Treon, S.; Munshi, N.; Richardson, P.; Anderson, K. (2002). Activation of NF-kappaB and upregulation of intracellular anti-apoptotic proteins via the IGF-1/Akt signaling in human multiple myeloma cells: therapeutic implications. *Oncogene*, **21**, 5673–5683.

37. Gnecchi, M.; He, H.; Noiseux, N.; Liang, O. D.; Zhang, L.; Morello, F.; Mu, H.; Melo, L. G.; Pratt, R. E.; Ingwall, J. S.; Dzau, V. J. (2006). Evidence supporting paracrine hypothesis for Akt-modified mesenchymal stem cell-mediated cardiac protection and functional improvement, *FASEB J.*, **20**, pp. 661–9.

38. Turing, A. M. (1952). The chemical basis of morphogenesis, *Philos. Trans. R. Soc. Lond. B Biol. Sci.*, **237**, pp. 37–72.

39. Bansagi, T.; Vanag, V. K.; Epstein, I. R. (2011). Tomography of reaction-diffusion microemulsions reveals three-dimensional Turing patterns, *Science*, **331**, pp. 1309–1312.

40. Sheth, R.; Marcon, L.; Bastida, M. F.; Junco, M.; Quintana, L.; Dahn, R.; Kmita, M.; Sharpe, J.; Ros, M. A. (2012). Hox genes regulate digit patterning by controlling the wavelength of a Turing-type mechanism, *Science*, **338**, pp. 1476–1480.

41. Kondo, S.; Miura, T. (2010) Reaction-diffusion model as a framework for understanding biological pattern formation, *Science*, **329**, pp. 1616–1620.

42. Garfinkel, A.; Tintut, Y.; Petrasek, D.; Bostrom, K.; Demer, L. L. (2004). Pattern formation by vascular mesenchymal cells, *Proc. Natl. Acad. Sci. USA*, **101**, pp. 9247–9250.

43. Epstein, I. R. (1991). Spiral waves in chemistry and biology, *Science*, **252**, pp. 67–67.

44. Vincent, S.; Perrimon, N. (2001). Developmental biology: Fishing for morphogens, *Nature*, **411**, pp. 533.

45. Eldar, A.; Dorfman, R.; Weiss, D.; Ashe, H.; Shilo, B. Z.; Barkai, N. (2002). Robustness of the BMP morphogen gradient in Drosophila embryonic patterning, *Nature*, **419**, pp. 304–308.

46. Yelon, D.; Stainier, D. Y. R. (2002). Pattern formation: Swimming in retinoic acid, *Curr. Biol.*, **12**, pp. R707–R709.

Chapter 3

Three-Dimensional Patterns of Tissues Emerging in Hydrogels

3.1 Background

In recent years, reliable therapies have been sought to regenerate damaged tissues and organs [1–4]. In order to achieve effective strategies to regenerate the functions of living tissues and organs, many cell-based methods have been proposed to heal or reconstitute and restore tissue functions [2, 5–7]. Tissue engineering based on complex and sophisticated scaffolds [8], cell sheet technology [9, 10], spatially manipulating living cells, or using a jet-based cell printer [11–13] is widely used to engineer tissues and organs. Yet, the structural complexity of engineered tissue is constrained by the mechanical precision of those approaches. Additionally, such artificial attempts need complex implementation procedures, and can cause immunogenic reactions during scaffold seeding and degradation [14] as well as influence cellular differentiation due to scaffold stiffness [15] and geometric cues [16]. To transcend these limitations, attention has turned to use self-organizing properties and the innate regenerative capability of the tissue/organism [17, 18], which can lead to the recapitulation of morphological structures

Self-Organized 3D Tissue Patterns: Fundamentals, Design, and Experiments
Xiaolu Zhu and Zheng Wang
Copyright © 2022 Jenny Stanford Publishing Pte. Ltd.
ISBN 978-981-4877-77-0 (Hardcover), 978-1-003-18039-5 (eBook)
www.jennystanford.com

and organizations of native tissue. In order to recapitulate the formation process of natural tissue via a self-assembly style, the cellular self-assembly behavior and its regulating method under certain conditions of extracellular matrices (ECM) and bio-signals should be studied.

However, the mechanism underlying tissue and organ formation via multicellular self-organization are not clear. Balanced organogenesis requires a complicated orchestration of cellular/environmental interactions and evolving cell behavior. Engineered three-dimensional (3D) hydrogel matrices [7, 19] provide a physiological context that could mimic natural ECM; thus events such as intercellular interactions [20] and self-assembly [21] can be recapitulated. The corresponding techniques include micro-patterning of hydrogels [20], nano-patterning techniques [22], establishing chemokine gradients in 3D matrices [23], culturing cell aggregates in 3D gel [24, 25], and obtaining liver organ buds by cellular self-organization in co-cultures of hepatic endoderm cells (HEs), human umbilical vein endothelial cells (HUVECs), and human mesenchymal stem cells (MSCs) [26]. Yet these methods need special extra control techniques or empirical protocols. The emergence of these structural morphologies in complex biological systems often cannot be predicted deterministically due to insufficient understanding of the underlying mechanism.

Previously, multipotent vascular mesenchymal cells (VMCs) have been shown to form 2D patterns in vitro [27–29], laying a foundation for its ability to form 3D patterns in vitro. Here, we investigated control methodologies for VMCs to form self-organized patterns in 3D. We experimentally demonstrate self-organized 3D-pattern formation (see Figure 3.1) in a modified hyaluronic acid (HA) hydrogel. By optimizing the components of the hydrogel and applying exogenous proteins, various 3D architectures consisting of aggregated VMCs can be formed in HA hydrogels. In these aggregates, network structures mimic the 3D features of trabecular bone; the multicellular spheroids serve as building blocks for organ regeneration [30]. These engineered 3D multicellular architectures are comparable with the predictions of the revised reaction-diffusion (RD) theoretical model described in this book.

3.2 Experimental Methods for 3D Culture of Cells

The experimental methods involved in 3D cell culture in hydrogels have some common protocols, but also vary when different hydrogels or cells involved. Here, we used the acrylated hyaluronic acid (HA-AC) hydrogel as our typical extracellular matrix (ECM). This hydrogel has several components and the proportion of each component can be tuned by researchers based on the specific consideration of their unique experiment design. The amount of the formed hydrogel does not require a large volume, but only 10–30 μL is needed in many practical cases. The readers may be suggested to make each hydrogel sample of low volume and fabricate more hydrogel samples. The adequate number of hydrogel samples will be used to complete more tests at once. For choosing the culture plates, the plate with small wells (such as 96–well plate) would be suggested because small wells not only provide a compact space for the small hydrogel samples, but also make the cytokines or growth factors secreted by cells, reaching the "working concentration" as soon as possible. These experiences can be validated just by comparing the cell growth situations in two wells added with different volumes of culture medium, such as 100 μL and 1 mL, if other conditions are all the same. The detailed experimental methods are described below. The descriptions include typical methods and materials used in the experiments for modifying the HA hydrogel and 3D culture in hydrogels.

3.2.1 Cell Culture

Vascular mesenchymal cells were isolated and cultured as described in literatures [31–33]. The cells were grown in Dulbecco's Modified Eagle's Medium (DMEM, Invitrogen) supplemented with 15% heat-inactivated fetal bovine serum (FBS) and 1% penicillin/ streptomycin (10,000 I.U./10,000 μg/mL; all from Mediatech, VA). The cells were incubated at 37 °C in a humidified incubator (5% CO_2 and 95% air), and were dissociated from the dish bottom using 0.25% trypsin-EDTA following standard protocols and passaged every three days.

3.2.2 HA Modification

HA-AC was prepared by using a two-step synthesis process. HA (60 kDa, Genzyme Corporation, Cambridge, MA) (2.0 g, 0.033 mmol) was reacted with adipic dihydrazide (ADH) (18.0 g, 105.5 mmol) at a pH of 4.75 in the presence of 1-ethyl-3-[3-dimethylaminopropyl] carbodiimide hydrochloride (EDC, 4.0 g, 20 mmol) overnight. The solution was purified through dialysis (8000 MWCO) in deionized water for two days. The purified intermediate (HA-ADH) was lyophilized and stored at −20 °C. HA-ADH (1.9 g) was re-suspended in 4-(2-hydroxyethyl)-1-piperazine ethane-sulfonic acid (HEPES) buffer (10 mM HEPES, 150 mM NaCl, 10 mM EDTA, pH 7.4) and reacted with N-acryloxysuccinimide (NHS-AC, 1.33 g, 4.4 mmol) overnight and purified through dialysis in DI water for two days before lyophilization. The final product was analyzed with 1H NMR (D_2-O) and the degree of acrylation (16%) determined by dividing the multiplet peak at $\delta = 6.2$ (cis and trans acrylate hydrogens) by the singlet peak at $\delta = 1.6$ (singlet peak of acetylmethyl protons in HA).

3.2.3 VMCs-Laden HA Hydrogel Formation

HA hydrogels were formed by Michael addition of bis-cysteine containing MMP-degradable crosslinker onto HA-AC pre-functionalized with cell adhesion peptides (RGD). 8 mg of lyophilized HA-AC was dissolved in 100 μL of 0.3 M TEOA buffer. Lyophilized aliquots of the RGD peptides (0.1 mg/vial) were dissolved in TEOA buffer and mixed with HA-AC solution, and finally kept for reacting for 25 min at 37 °C to get the HA-RGD solution. The cells were dissociated from the petri dish bottom by 0.25% trypsin-EDTA, and the resuspended cell solution was placed on ice. A lyophilized aliquot of the crosslinker (1.0 mg) was then diluted in 20 μL of 0.3 M TEOA buffer (pH = 8.4) immediately before mixing with HA-AC solution, HA-RGD (final concentration of 100–150 μM RGD), and the cell solution. The gel precursor solution was pipetted onto a sigmacoted glass slide in drop-wise (10 μL per drop), then clamped with another sigmacoted slide with plastic cover-slip spacers, and finally incubated for 30 min at 37 °C to allow for gelation. The final gel was swelled in culture media before being placed inside 96-well plates for the following long-term culture. The mechanical properties of the hydrogels can be controlled by varying the HA concentration or the crosslinking density, r, which is defined as mole ratio of thiol (-SH) from the crosslinkers to acrylate group (-ACs) from the HA-ACs.

3.2.4 Fixation and Fluorescent Staining

Before fixing the cells inside the hydrogel, the gel samples were washed twice in the micro-well plate with prewarmed phosphate-buffered saline (PBS). Then the cells in gel were fixed by 4% paraformaldehyde (PFA) solution for 40 min at room temperature. The samples were stained by fluorescent dyes that can render the F-actins (Alexa Fluor 488 phalloidin, Invitrogen) and the nuclei (DAPI, Invitrogen) inside the cells. For F-actin staining, 1% bovine serum albumin (BSA) was added to the staining solution of Alexa Fluor 488 phalloidin to reduce nonspecific background staining, and then the staining solution was pipetted to cover the cell-laden hydrogel sample in microwells for 25 min at room temperature. For nuclei staining, the solution of 300 nM DAPI in PBS was added to microwells containing the cell-laden hydrogel samples and the sample was incubated for 5 min.

3.2.5 Clustered Encapsulation of Cells in 3D HA Hydrogels

Clustered encapsulation of cells was made for studying cellular migration in HA hydrogels. Cell clusters in fibrin gel clots were made by resuspending 300,000 VMCs in 10 μL of fibrin and thrombin solution (2 mg/mL fibrinogen and 2 U/mL thrombin). Then, the clusters of cells were made by dropping the suspension onto a sigmacoted plate and incubating at 37 °C for 20 min. The clusters of cells inside fibrin clots were transferred into the HA hydrogels by placing it inside the gel precursor solution. The gel was swelled in DMEM and cultured in DMEM supplemented with 15% FBS and 1% P/S.

3.2.6 3D Visualization Using Selective Plane Illumination Microscopy (SPIM)

After encapsulating in the hydrogel samples, cells were stained by the phalloidin and DAPI. The HA-gel samples were immersed into a low gelling-temperature agarose solution (37.5°C) which was composed of 0.5% (w/w) agarose and the rest of PBS. The agarose was gelled at room temperature and made the HA-gel sample mounted inside it. Then, the cuvette was placed onto the automatic stage of the SPIM platform. The sample was optically sliced when the stage was programmed to travel at 1 μm or 2 μm per step. The SPIM imaging platform is more suitable for observing samples in large vision field using 4′ or 10′ objectives, which usually has higher imaging quality than common confocal microscopy. The obtained image sequences were stacked and processed by commercial 3D reconstruction software (Amira 5.2 Resolve RT, trial version). The basic principle of SPIM is stated in the reference [30] in Chapter 1.

3.3 Results and Discussion for 3D Microtissue Patterns Emerging in Ha Hydrogels

The cell populations can grow in the modified HA hydrogel, and the cells can sprout, spread, proliferate, and connect with other cells in 3D hydrogel matrices.

3.3.1 3D Pattern Formation of VMCs in Modified HA Hydrogel

As shown in Figure 3.1a, the VMCs were encapsulated and cultured inside the modified HA hydrogel. Partial carboxyl groups in the HA backbone were modified with acrylate groups to incorporate RGD peptides, to introduce integrin binding sites and to crosslink the HA polymers, to generate a mechanically stable hydrogel, through Michael addition chemistry. Cells attached to the RGD peptides and distributed in a 3D space after the hydrogel were gelled. Then the cell-encapsulated HA hydrogels were cultured in microwells in cell culture plates as shown in Figure 3.1b. The RGD concentration influenced both the starting time and the extent of cell spreading. In general, with high concentration of RGD, cells spread earlier, and migrated faster. However, high concentration of RGD gave rise to low proliferation rate. The HA concentration and crosslinker density determine the hydrogel stiffness and influence the cell behavior as well. Cells in hydrogels with lower stiffness showed more spreading, migration, and slower proliferation rate, but they presented a higher degradation rate of hydrogel not suitable for long-term cell culture. Therefore, the components in this semi-synthetic HA hydrogel were optimized to provide an effective platform to study multicellular structure formation in a 3D matrix. In this study, the HA hydrogels with 3.0–3.5% HA, 100 –150 µM RGD, and thiol to acrylate ratio (r ratio) values ranging from 0.4 to 0.75 were chosen as the optimized conditions that facilitate multicellular structure formation. The VMC is a stem cell-like multipotent cell that exhibits several remarkable capabilities for multicellular structure formation, which was studied in our previous work [28, 34] and other fundamental work [29, 35].

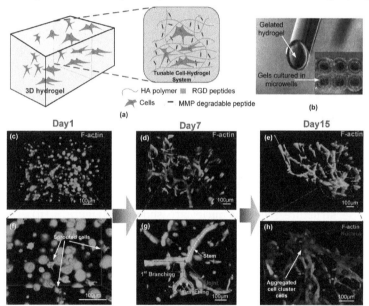

Figure 3.1 The capability of VMCs to form patterns in 3D HA hydrogel. (a) The schematics for 3D cell culture inside semi-synthetic modified HA hydrogel. (b) The gelated HA hydrogels encapsulating cells are cultured in microwells. In 3D culture, VMCs nearly uniformly distributed in 3D HA hydrogel at day1 (24 h after cell seeding) and only a few cells sprouted (c, f). On day 7, connected cells formed a network with branched structures (d, g). On day 15, local cells aggregated into thicker and bulkier structure (e, h). The cells were stained by fluorescent dyes, which can render the F-actin (green color) and the nuclei (blue color) inside the cells. Scale bars, 100 µm.

The VMCs initially presented a uniform distribution. The cells can spread, migrate, and proliferate inside the modified HA hydrogel. Figure 3.1c shows randomly distributed round cells inside the 3D hydrogel one day after cell seeding. A few cells had small sprouting projections. After 2–3 days, the VMCs apparently spread out with finger-like projections and usually kept this morphology in the following week. VMCs started remarkable interaction with neighboring cells as early as day 2, which became more pronounced between days 4 and 8. 3D multicellular networks formed, consisted of a large number of branches and sub-branches (Figure 3.1d). Each branch and sub-branch was formed by local multicellular aggregation. In the branching process, there was no specific order in which the stem, branches and sub-branches (Figure 3.1g) appear according to the continuous observation during the culture process. These three parts (stem, branches, and sub-branches) seemed to emerge at the same time, aggregate respectively in local regions, and then connect with each other under the combined stimuli of various physical and chemical signals in 3D. Furthermore, the plane in which the second branching occurs is almost perpendicular to the plane in which the first branching lays, as presented in Figure 3.1g. On day 15, local cells aggregated into a thicker and bulkier structure as shown in Figures 3.1e and h. Gradually, many cells aggregated and connected with each other and finally formed a large multicellular dense structure (cellular nuclei are shown in Figure 3.1h). This dense structure had an irregular and rough appearance in the 3D matrix. The aggregated multicellular structure had a feature size of more than 200 μm (Figure 3.1h). It has much greater cell density and longitudinal dimension compared to those of single branches as presented in Figures 3.1d and g. After day 15, the multicellular structures displayed no significant variation until the hydrogel showed obvious degradation.

3.3.2 Generation of Varying Morphologies of 3D Structures Composed of Aggregated Cells

We have demonstrated that the morphologies of 3D structures composed of cell populations can be controlled by varying the global context of the cultured samples. Specifically, under the action of different uniformly distributed exogenous factors, three types of morphologies of 3D structure consisting of aggregated cells were created. Cells inside the hydrogel can sensitively respond to the exogenous factors. Although these factors (proteins) influenced the biophysical condition of each individual cell equally, the kinetic responses of cells in different regions were distinct according to the observed multicellular morphology of the cultured samples. Figure 3.2a shows the honeycombed network structure consisting of around one million cells. The cells are closely connected with each other and have local aggregations, but still leave hollow spaces between the connecting links. At the edge of this gel sample, many more cells formed an aggregated smooth bossing, which has many cells branching out. As shown in Figures 3.2a and d, connecting links exist among both multicellular aggregates and individual cells, and these links go through the 3D space with either curved or straight geometries in a nearly isotropic manner. Figure 3.2b shows a more aggregated multicellular structure with similar bough geometry when Noggin (900 ng/mL, the inhibitor of bone morphogenetic protein-4 [BMP-4]) was applied into the 3D matrix. This structure has local thick aggregates of cells and filament-like connections among these separated aggregates. The enlarged morphology from a different viewing angle is presented in Figure 3.2e. Its aggregated lump has a dimension larger than 200 µm. The connecting filament-like cellular structures are mostly perpendicular to the edges of this lump. More remarkably, spheroids with diameters of around 200 µm formed in around 13 days after the exogenous BMP-2 (500 ng/mL, activator of Matrix Gla Protein [MGP]) was applied as shown in Figures 3.2c and f. There were about a dozen spheroids in hydrogel-based culture inside each of the micro-wells of the culture plate. Each spheroid was packed with thousands of cells. Most of the spheroids had connections with adjacent ones (Figure 3.2c). More details on the evolution of the network or spheroids composed of numerous cells can be found in the literature [36].

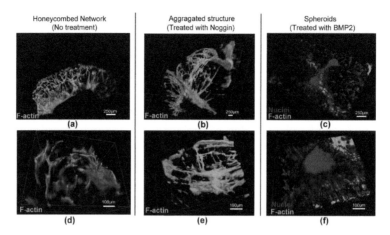

Figure 3.2 Uniformly distributed protein treatments alter the 3D self-organized multicellular structure composed of aggregates of VMCs from networks (a) to thick bough-like aggregates (b), then to spheroids (c), (d) shows the zoom-in of (a), (b), and (e) that present the multicellular structure treated by Noggin (Day 11), and (c) and (f) present the multicellular structure treated by BMP-2 (Day 13). The F-actin inside cells was stained with Alexa Fluor 488 phalloidin, which could outline the cellular skeletons. The nuclei of cells in (c) and (f) were stained with DAPI. The conditions for making HA hydrogel: 3.5%HA, 100 μM RGD, initial cell density = 7500/μL.

We proposed that the observed morphologies of 3D multicellular architectures are associated with a Turing-type mechanism. The mechanism has been used to explain the pattern formation in nature (such as patterns on the skins of tigers), but it has not been proved to be capable of interpreting and guiding the 3D multicellular architecture formation in vitro. As first proposed by Turing [37], biological patterns emerging from instability of a homogeneous steady state in biology can often be modeled mathematically by postulating "morphogens" that react chemically and also diffuse. The evolution of complex multicellular patterns can be interpreted by the RD framework, in which chemicals produced by cells interact as activators and inhibitors (morphogen pairs) and diffuse through the matrices at distinct rates. The RD equations modeling morphogen pairs have been used to analyze pattern formation in many chemical and biological systems [29, 38–41]. The existence of morphogens and their effects on tissues in vertebrates have been proved [42]. Specific morphogens for creating biological patterns have been identified, including transforming growth factor (TGF) β family members, proteins of the Wnt and Hedgehog families, and retinoic acid [39, 42–44]. Certain genes modulate the molecular signaling of morphogens [39], which presents a more complicated working mechanism of reaction and diffusion of morphogens.

BMP-2 and BMP-4, members of the TGF superfamily, and their respective inhibitors, such as matrix gammacarboxyglutamic acid protein (MGP), have already been identified as morphogens that are secreted by VMCs [29]. Here, we treat the two activators as equivalent, thus our variables model combined the effects of these two activators. We modeled our cell-hydrogel system consisting of slowly-diffusing activator with autocatalytic reaction (using Gierer and Meinhardt kinetics), its rapidly-diffusing inhibitor, and cell density, reflecting proliferation, cytokinetic motility, and chemotactic migration with respect to activators, as functions of a 3D domain (x, y, z) in dimensionless forms:

$$\frac{\partial U}{\partial t} = D\nabla^2 U + \gamma \left[\frac{pnU^2}{V\left(1 + kU^2\right)} - cU \right] + ExU \tag{3.1}$$

$$\frac{\partial V}{\partial t} = \nabla^2 V + \gamma \left[bnU^2 - eV \right] + ExV \tag{3.2}$$

$$\frac{\partial n}{\partial t} = q\nabla^2 n - \chi \left[\nabla \cdot \left(\frac{n}{(k_n + U)^2} \nabla U \right) \right] + r_n n(1 - n) \tag{3.3}$$

In Eqs. (3.1)–(3.3), U, V, and n are dimensionless concentrations of activator (U), inhibitor (V), and cells (n) as functions of space coordinate (x, y, z) and time (t); $D\nabla^2 U$, $\nabla^2 V$, and $q\nabla^2 n$ are three analogical terms to describe the diffusion of activators, inhibitors, and cells. c and e are the degradation rates of the activator and inhibitor; b is the coefficient representing relative production of inhibitor to activator; and χ is the regulation factor for chemotactic migration in response to the gradient of activators. D and q are the ratios of diffusion coefficients for activator-to-inhibitor, and cells-to-inhibitor, respectively. γ is a scaling factor related to domain size, biosynthetic timescale, and inhibitor diffusivity. ExU and ExV are the exogenous source terms for activator and inhibitor, respectively. The values for the group of parameters were set empirically by considering both the experimental setup and the intrinsic law of the Turing RD theory.

Network	Thick aggregations	Spheroids

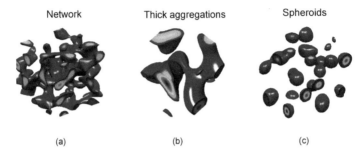

(a) (b) (c)

Figure 3.3 (a–c) Computational simulations showing n (x, y) as (a) a honeycombed networks structure $(D = 0.005, q = 0.004, \chi = 0.09, ExU = 0, ExV = 0)$, (b) aggregated bough-shaped structure $(D = 0.005, q = 0.007, \chi = 0.07, ExU = -5 \times 10^{-5}, ExV = 0)$, and (c) distributed spheroids $(D = 0.005, q = 0.004, \chi = 0.1, ExU = 5 \times 10^{-6}, ExV = 0)$. The volumes of 3D simulation regions are 1 mm^3. The parameter values have different values from the publications [36] because these parameters can be set with different groups of values as long as the assumptions for parameter values are reasonable and could lead to a target 3D morphology. The relative magnitudes among a group of parameters should qualitatively be in accordance with the designed experimental conditions.

The simulation results are shown in Figure 3.3, which are similar to the experimental results in Figure 3.2. The exogenous Noggin (50 kDa, monomer) has a slower diffusion rate than BMP-2 and BMP-4. Thus, Noggin does not satisfy the criteria that the inhibitor must diffuse more rapidly than the activator in Turing-type activator-inhibitor models. Although Noggin cannot serve as an inhibitor in this model, it still regulates the inhibitory process of MGP to BMP-2/4 since the Noggin protein binds BMP-4 with high affinity and can abolish BMP-4 activity [45]. Therefore, the BMP-4 was somewhat reduced, and the ExU was correspondingly assumed as negative value (-5×10^{-5}). In contrast, the ExU was set as positive value (5×10^{-6}) when exogenous BMP-2 was applied.

The migration rate of cells is determined by the net results of the cellular diffusion and chemotactic effect, which counteract each other for cellular motions. In the simulation model, the cellular diffusion of cells could be enhanced by increasing the value of q, and the chemotactic motion of cells could be enhanced by increasing the value of χ. According to the migration tests for cell clusters inside HA hydrogels with the above experimental conditions (see Figure 3.4), the migration rate of the cells in Noggin-treated sample is larger than that in the control (normal sample). Thus, the $q = 0.007$ and $\chi = 0.07$ for Noggin-treated sample is set compared to $q = 0.004$ and $\chi = 0.09$ for the control, since larger q and smaller χ lead to a quicker migration away from an aggregated cellular cluster. On the other hand, the experimentally measured migration rate of the cells in BMP-2-treated sample is smaller than that in the control after day 2. Thus, the increased $\chi = 0.1$ for BMP-2-treated sample is set compared to the control, since larger χ value makes the cluster of cells more aggregated, which leads to a slower migration away from the cellular cluster. All other parameter values are the same among these three cases. In summary, the mathematical model based on Turing-type mechanism can interpret the switch among different morphological regimes of 3D multicellular architectures when different exogenous factors are applied.

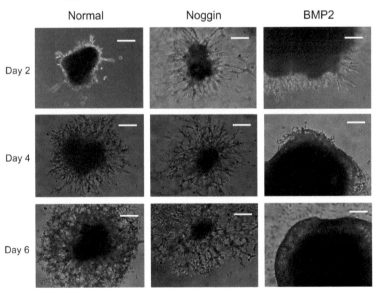

Figure 3.4 Cellular migration test for VMC culture treated by exogenous proteins. Noggin made the cells migrate out of the cellular cluster in fibrin clot more quickly than the control. BMP-2 made the cells migrate out of the clot a bit quicker with first two days than control, but conspicuously slower than the control after day 2. After day 3, the VMC culture treated by BMP-2 had a slower migration rate than both the control and Noggin-treated VMCs. Conditions: 3.5% HA, 100 μM RGD, $r = 0.6$. All the scale bars, 200 μm.

3.3.3 Mapping Combined Effects of Exogenous Factors

The regulation of 3D multicellular architectures described above depends on applying single exogenous factors, and other determinants, such as culture time, are not taken into account as a modulation option. An extended method with more degrees of freedom for holistically regulating the 3D multicellular structures is proposed here. In this more generalized method, exogenous factors singly and in combination are applied to obtain a variety of morphologies at different stages of the process. These morphologies are mapped to the three axes corresponding to the concentration of BMP-2, concentration of Noggin and culture duration, respectively, as presented in Figure 3.5. It shows that the cells form spheroids when BMP-2 is solely applied. More spheroids, having slightly bigger size, emerged with longer duration of the culture (positive direction in z-axis). When Noggin is applied singly, the cells usually form local aggregates, and then form small lumps with irregular geometry. The small lumps become denser and bigger after day 14. More interestingly, the combinational effect of the two factors (BMP-2 and Noggin) leads to hybrid multicellular morphologies, in which the distinct morphological features respectively induced by exogenous BMP-2 and Noggin, can coexist in the same sample. For example, the morphology under the action of combined factors having the features of both lumps and spheroids emerged at day 14, but the geometrical size and density of the lumps and spheroids are usually different from that in the cultures treated by corresponding single factors. Finally, these multicellular structures with combinational features will grow further and have distinct, more aggregated and complex topologies. More types of 3D multicellular structures can form under the combined impact of these three factors. Mapping the combined effects of the various exogenous factors lays a foundation for promoting the capability and flexibility of the creation of multicellular structures with numerous different morphologies.

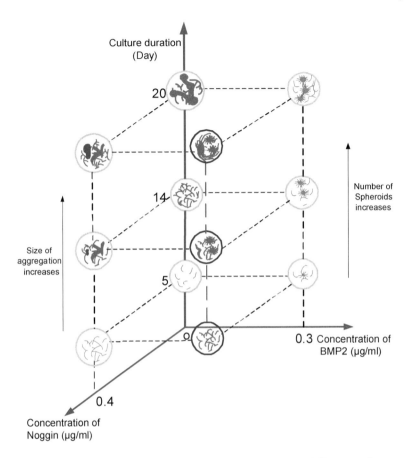

Figure 3.5 The effects on multicellular morphology at different culture date by sole and combinational exogenous factors. The experimental conditions for making these gel samples: 3.5% HA, 100 μM RGD, $r = 0.55$.

3.4 Summary

A simulation model based on Turing-type mechanism is built for simulating cellular self-organization in 3D hydrogels with inner interfaces. The simulation results show the self-organized tubular and spherical structures with inner cavities that can be generated around the interfacial planes inside the 3D matrices. This simulation model mathematically describes suitably-tuned RD system with the recognized activator (bone morphogenic protein-2) and its inhibitor (matrix GLA protein) for VMCs. The generated hollow structures self-organized by cells are robust and can sustain for a long time in this dynamic simulation system. Our simulation model can rationally predict 3D multicellular tubular structure formation in a heterogeneous matrix, which lays a foundation for regenerating injured vasculatures in organs, such as hearts, lungs, and kidneys, more efficiently in a rational manner.

Acknowledgments

This work is supported by the National Natural Science Foundation of China (Grant No. 51505127) and Fundamental Research Funds for the Central Universities (Grant Nos. 2018B22414 and 2015B04414) of China.

References

1. Mimeault, M.; Batra, S. K. (2006). Concise review: Recent advances on the significance of stem cells in tissue regeneration and cancer therapies, *Stem Cells,* **24**, pp. 2319–45.

2. Eiraku, M.; Takata, N.; Ishibashi, H.; Kawada, M.; Sakakura, E.; Okuda, S.; Sekiguchi, K.; Adachi, T.; Sasai, Y. (2011). Self-organizing optic-cup morphogenesis in three-dimensional culture, *Nature,* **472**, pp. 51–6.

3. Rizzoti, K.; Lovell-Badge, R. (2011). Regenerative medicine: Organ recital in a dish, *Nature,* **480**, pp. 44–6.

4. Bashur, C. A.; Venkataraman, L.; Ramamurthi, A. (2012). Tissue engineering and regenerative strategies to replicate biocomplexity of vascular elastic matrix assembly, *Tissue Engineering Part B: Reviews,* **18**, pp. 203–217.

5. Kim, S.; Shah, S. B.; Graney, P. L.; Singh, A. (2019) Multiscale engineering of immune cells and lymphoid organs. *Nature Reviews Materials,* **4**, 355–378.

6. Berthiaume, F.; Maguire, T. J.; Yarmush, M. L. (2011). Tissue engineering and regenerative medicine: History, progress, and challenges, *Annu Rev Chem Biomol Eng,* **2**, pp. 403–430.

7.   Haycock, J. W. (2011). 3D cell culture: A review of current approaches and techniques, *Methods Mol Biol*, **695**, pp. 1–15.

8.   Tsang, V. L.; Bhatia, S. N. (2004). Three-dimensional tissue fabrication, *Adv Drug Deliv Rev*, **56**, pp. 1635–47.

9.   Shimizu, T.; Yamato, M.; Kikuchi, A.; Okano, T. (2003). Cell sheet engineering for myocardial tissue reconstruction, *Biomaterials*, **24**, pp. 2309–16.

10. Yang, J.; Yamato, M.; Shimizu, T.; Sekine, H.; Ohashi, K.; Kanzaki, M.; Ohki, T.; Nishida, K.; Okano, T. (2007). Reconstruction of functional tissues with cell sheet engineering, *Biomaterials*, **28**, pp. 5033–43.

11. Odde, D. J.; Renn, M. J. (2000) Laser-guided direct writing of living cells. *Biotechnol Bioeng*, **67**, pp. 312–8.

12. Mironov, V.; Boland, T.; Trusk, T.; Forgacs, G.; Markwald, R. R. (2003). Organ printing: Computer-aided jet-based 3D tissue engineering, *Trends Biotechnol*, **21**, pp. 157–61.

13. Ho, C.-T.; Lin, R.-Z.; Chang, W.-Y.; Chang, H.-Y.; Liu, C.-H. (2006). Rapid heterogeneous liver-cell on-chip patterning via the enhanced field-induced dielectrophoresis trap, *Lab on a Chip*, **6**, pp. 724–734.

14. Jakab, K.; Norotte, C.; Marga, F.; Murphy, K.; Vunjak-Novakovic, G.; Forgacs, G. (2010). Tissue engineering by self-assembly and bio-printing of living cells, *Biofabrication*, **2**, pp. 022001.

15. Levy-Mishali, M.; Zoldan, J.; Levenberg, S. (2009). Effect of scaffold stiffness on myoblast differentiation, *Tissue Engineering Part A* **15**, pp. 935–944.

16. Kilian, K. A.; Bugarija, B.; Lahn, B. T.; Mrksich, M. (2010). Geometric cues for directing the differentiation of mesenchymal stem cells, *Proc. Natl. Acad. Sci. USA*, **107**, pp. 4872–4877.

17. Mironov, V.; Trusk, T.; Kasyanov, V.; Little, S.; Swaja, R.; Markwald, R. (2009). Biofabrication: A 21st century manufacturing paradigm, *Biofabrication*, **1**, 022001:1–16.

18. Lei, Y.; Gojgini, S.; Lam, J.; Segura, T. (2011). The spreading, migration and proliferation of mouse mesenchymal stem cells cultured inside hyaluronic acid hydrogels, *Biomaterials*, **32**, pp. 39–47.

19. Lee, J.; Cuddihy, M. J.; Kotov, N. A. (2008). Three-dimensional cell culture matrices: State of the art, *Tissue Engineering Part B-Reviews*, **14**, pp. 61–86.

20. Trkov, S.; Eng, G.; Di Liddo, R.; Parnigotto, P. P.; Vunjak-Novakovic, G. (2010). Micropatterned three-dimensional hydrogel system to study human endothelial-mesenchymal stem cell interactions, *J. Tissue Eng. Regen. Med.*, **4**, pp. 205–215.

21. Jakab, K.; Norotte, C.; Damon, B.; Marga, F.; Neagu, A.; Besch-Williford, C. L.; Kachurin, A.; Church, K. H.; Park, H.; Mironov, V.; Markwald, R.; Vunjak-Novakovic, G.; Forgacs, G. (2008). Tissue engineering by self-assembly of cells printed into topologically defined structures, *Tissue Eng Part A,* **14,** pp. 413–421.

22. Griffin, D. R.; Borrajo, J.; Soon, A.; Acosta-Vélez, G. F.; Oshita, V.; Darling, N.; Mack, J.; Barker, T.; Iruela-Arispe, M. L.; Segura, T. (2014). Hybrid photopatterned enzymatic reaction (HyPER) for in situ cell manipulation, *ChemBioChem,* **15,** 233–242.

23. Haessler, U.; Pisano, M.; Wu, M.; Swartz, M. A. (2011). Dendritic cell chemotaxis in 3D under defined chemokine gradients reveals differential response to ligands CCL21 and CCL19, *Proc Natl Acad Sci USA,* **108,** pp. 5614–9.

24. Eiraku, M.; Takata, N.; Ishibashi, H.; Kawada, M.; Sakakura, E.; Okuda, S.; Sekiguchi, K.; Adachi, T.; Sasai, Y. (2011). Self-organizing optic-cup morphogenesis in three-dimensional culture, *Nature,* **472,** pp. 51–56.

25. Suga, H.; Kadoshima, T.; Minaguchi, M.; Ohgushi, M.; Soen, M.; Nakano, T.; Takata, N.; Wataya, T.; Muguruma, K.; Miyoshi, H.; Yonemura, S.; Oiso, Y.; Sasai, Y. (2011). Self-formation of functional adenohypophysis in three-dimensional culture, *Nature,* **480,** pp. 57–62.

26. Takebe, T.; Sekine, K.; Enomura, M.; Koike, H.; Kimura, M.; Ogaeri, T.; Zhang, R. R.; Ueno, Y.; Zheng, Y. W.; Koike, N.; Aoyama, S.; Adachi, Y.; Taniguchi, H. (2013). Vascularized and functional human liver from an iPSC-derived organ bud transplant, *Nature,* **499,** pp. 481–484.

27. Chen, T. H.; Hsu, J. J.; Zhao, X.; Guo, C. Y.; Wong, M. N.; Huang, Y.; Li, Z. W.; Garfinkel, A.; Ho, C. M.; Tintut, Y.; Demer, L. L. (2012). Left-right symmetry breaking in tissue morphogenesis via cytoskeletal mechanics, *Circul. Res.,* **110,** pp. 551-559.

28. Chen, T. H.; Zhu, X.; Pan, L.; Zeng, X.; Garfinkel, A.; Tintut, Y.; Demer, L. L.; Zhao, X.; Ho, C. M. (2012). Directing tissue morphogenesis via self-assembly of vascular mesenchymal cells, *Biomaterials,* **33,** pp. 9019-26.

29. Garfinkel, A.; Tintut, Y.; Petrasek, D.; Bostrom, K.; Demer, L. L. (2004). Pattern formation by vascular mesenchymal cells, *Proc. Natl. Acad. Sci. USA,* **101,** pp. 9247–9250.

30. Mironov, V.; Visconti, R. P.; Kasyanov, V.; Forgacs, G.; Drake, C. J.; Markwald, R. R. (2009). Organ printing: Tissue spheroids as building blocks, *Biomaterials,* **30,** pp. 2164–2174.

31. Bostrom, K.; Watson, K. E.; Horn, S.; Wortham, C.; Herman, I. M.; Demer, L. L. (1993). Bone morphogenetic protein expression in human atherosclerotic lesions, *J. Clin. Invest.,* **91,** pp. 1800–1809.

32. Hsiai, T. K.; Cho, S. K.; Reddy, S.; Hama, S.; Navab, M.; Demer, L. L.; Honda, H. M.; Ho, C. M. (2001). Pulsatile flow regulates monocyte adhesion to oxidized lipid-induced endothelial cells, *Arterioscler. Thromb. Vasc. Biol.,* **21**, pp. 1770–1776.

33. Wong, P. K.; Yu, F. Q.; Shahangian, A.; Cheng, G. H.; Sun, R.; Ho, C. M. (2008). Closed-loop control of cellular functions using combinatory drugs guided by a stochastic search algorithm, *Proc. Natl. Acad. Sci. USA,* **105**, pp. 5105–5110.

34. Chen, T. H.; Hsu, J. J.; Zhao, X.; Guo, C.; Wong, M. N.; Huang, Y.; Li, Z.; Garfinkel, A.; Ho, C. M.; Tintut, Y.; Demer, L. L. (2012). Left-right symmetry breaking in tissue morphogenesis via cytoskeletal mechanics, *Circ Res.,* **110**, pp. 551–9.

35. Abedin, M.; Tintut, Y.; Demer, L. L. (2004). Mesenchymal stem cells and the artery wall, *Circul. Res.,* 95, pp. 671–6.

36. Zhu, X.; Gojgini, S.; Chen, T. H.; Fei, P.; Dong, S.; Ho, C.-M.; Segura, T. (2017). Directing three-dimensional multicellular morphogenesis by self-organization of vascular mesenchymal cells in hyaluronic acid hydrogels, *Journal of Biological Engineering,* **11**, pp. 12.

37. Turing, A. M. (1952). The chemical basis of morphogenesis, *Philos. Trans. R. Soc. Lond. B Biol. Sci.,* **237**, pp. 37–72.

38. Bansagi, T.; Vanag, V. K.; Epstein, I. R. (2011). Tomography of reaction-diffusion microemulsions reveals three-dimensional Turing patterns, *Science,* **331**, pp. 1309–1312.

39. Sheth, R.; Marcon, L.; Bastida, M. F.; Junco, M.; Quintana, L.; Dahn, R.; Kmita, M.; Sharpe, J.; Ros, M. A. (2012). Hox genes regulate digit patterning by controlling the wavelength of a Turing-type mechanism, *Science,* **338**, pp. 1476–1480.

40. Kondo, S.; Miura, T. (2010) Reaction-diffusion model as a framework for understanding biological pattern formation, *Science,* **329**, pp. 1616–1620.

41. Epstein, I. R. (1991). Spiral waves in chemistry and biology, *Science,* **252**, pp. 67–67.

42. Vincent, S.; Perrimon, N. (2001). Developmental biology: Fishing for morphogens, *Nature,* **411**, pp. 533.

43. Eldar, A.; Dorfman, R.; Weiss, D.; Ashe, H.; Shilo, B. Z.; Barkai, N. (2002). Robustness of the BMP morphogen gradient in Drosophila embryonic patterning, *Nature,* **419**, pp. 304–308.

44. Yelon, D.; Stainier, D. Y. R. (2002). Pattern formation: Swimming in retinoic acid, *Curr. Biol.,* **12**, pp. R707–R709.

45. Zimmerman, L. B.; DeJesusEscobar, J. M.; Harland, R. M. (1996). The Spemann organizer signal noggin binds and inactivates bone morphogenetic protein 4, *Cell,* **86**, pp. 599–606.

Chapter 4

Constructing 3D Tissue Structures via Cellular Self-Assembly at Patterned Interfaces inside Hydrogel

4.1 Background

Nowadays, therapies for regenerating damaged biological tissues or organs attract more and more attentions [1–4]. In order to develop efficient strategies to regenerate the functions of living tissues and organs, many different cell-based therapeutic methods have been proposed to heal, reconstitute, or restore tissue functions [2, 5–7]. Scaffolds with increasing complexity and sophistication are widely used to serve as engineered extracellular matrix (ECM). To reconstitute tissue architectural features in microenvironments, diverse research attempts have been made to fabricate the scaffold with specific structure to guide cell spreading [8], assemble layers of cultured cell sheets [9, 10], directly deposit cells or move cells to chosen locations [11–13]. However, no matter how sophisticated scaffolds are, they can cause problems arising from their degradation, eliciting immunogenic reactions, and other priori unforeseen complications [14]. It is also being realized that ultimately the best approach might be to rely on the self-assembly and self-organizing properties of cells and tissues and the innate regenerative capability of the organism itself.

Self-Organized 3D Tissue Patterns: Fundamentals, Design, and Experiments
Xiaolu Zhu and Zheng Wang
Copyright © 2022 Jenny Stanford Publishing Pte. Ltd.
ISBN 978-981-4877-77-0 (Hardcover), 978-1-003-18039-5 (eBook)
www.jennystanford.com

As we know, the human body is self-assembled by numerous hierarchical tissue architectures in a natural style, and each type of tissue is hierarchically self-assembled by modular building blocks across multiple length scales [15], such as the modular and hierarchical organization of the human mammary gland [16]. In addition to the biological architectures inside the tissue, the tissue-to-tissue interfaces are also ubiquitous in human body and exhibit a gradient of structural and mechanical properties that serve a number of functions, from mediating load transfer between two distinct types of tissue to sustaining the heterotypic cellular communications required for interface function and homeostasis [17, 18]. Therefore, for recapitulating the architectural features of natural tissue in vitro via a self-assembly style, the cellular self-assembly behavior around the interface of different types of ECM should be studied. To study the mechanism and control approaches of the cellular self-organization at the interface of tissue architectural features in vitro, the cells need to be cultured in a suitable physiological environment that can mimic the natural ECM in our body. Currently, three-dimensional (3D) cell culture [7, 19] in hydrogels is a promising strategy to achieve this goal, since it provides the physiological environments more close to nature and takes full advantage of the inherent behaviors of the cells, such as the cellular interactions [20] or self-organization [21] in 3D matrix. The ECM is a non-cellular material, and a complex network of chemically different macromolecules (e.g. polysaccharides and proteins) that offers not only structural/mechanical support, but also essential chemical and physical guidance cues for cells [22]. Researchers have studied the in vitro effect or functions of various types of interface or the induced gradient by it in hydrogel systems. Simona et al. investigated the density gradients at hydrogel interfaces for enhanced cell penetration [23]; Kim et al. used a simple gradual freezing-thawing method to creating stiffness gradient polyvinyl alcohol hydrogel in order to investigate stem cell differentiation behaviors [24]; and Rao et al. demonstrated the inherent interfacial mechanical gradients in 3D hydrogels influence tumor cell behaviors [25]. Since the interfacial gradients influence various cell behaviors, the concentration gradient of cells/factors and structural gradients in hydrogel system have the possibility to result in morphology change of multicellular structure formed in hydrogel system, when the dynamic distribution of mechanical, chemical, and biological signals is changed in it.

Here we demonstrate a method that involves an engineered composite hydrogel structure including the stiffness interfaces inside it and steep gradient of cell concentration around the stiffness interfaces to serve as the 3D culture matrix for inducing the formation of 3D multicellular structures. The hyaluronic acid (HA) is used here because of its high biocompatibility, low immunogenicity [26, 27], and suitability for culturing stem cells. Semisynthetic HA hydrogels can be degraded by hyaluronidases and matrix metalloproteinases (MMPs) [28] and during the growing of cells. In addition, vascular mesenchymal cells (VMCs) are cultured in the upper part of this composite hydrogel structure to recapitulate the hierarchical tissue architectures and the heterogeneity of natural tissues.

4.2 Materials and Methods

4.2.1 Cell Culture

Vascular mesenchymal cells and bovine vascular endothelial cells (BVECs) were isolated and cultured as described [27, 29–31]. All cells were grown in Dulbecco's Modified Eagle's Medium (DMEM, Invitrogen) supplemented with 15% heat-inactivated fetal bovine serum and 1% penicillin/streptomycin (10,000 I.U./10,000 µg/mL; all from Mediatech, VA). The cells were incubated at 37 °C in a humidified incubator (5% CO_2 and 95% air), and were split using trypsin following standard protocols and passaged every 3 days.

4.2.2 HA Modification

Acrylated hyaluronic acid (HA-AC) was prepared by using a two-step synthesis. HA (1.0 g, 0.017 mmol, 60 kDa) was reacted with 18.0 g (105.5 mmol) adipic dihydrazide (ADH) at a pH of 4.75 in the presence of 2.0 g (10.41 mmol) 1-ethyl-3-[3-dimethylaminopropyl] carbodiimide hydrochloride (EDC) overnight and purified through dialysis (8000 MWCO) in DI water for a week. The purified intermediate (HA-ADH) was lyophilized and stored at −20 °C until used. Around 38.8% of the carboxyl groups were modified with ADH based on the trinitrobenzene sulfonic acid (TNBSA, Pierce, Rockford, Illinois) assay. HA-ADH (1.0 g, 0.014 mmol) was reacted with N-Acryloxysuccinimide (NHS-AC) (0.75 g, 4.4 mmol) in HEPES buffer (pH 7.2) overnight and purified through dialysis in DI water for a week before lyophilization. All the primary amines were acrylated based on the TNBSA assay.

4.2.3 HA Hydrogel Synthesis

HA hydrogels were formed by Michael addition of bis-cysteine containing MMP-degradable crosslinker onto HA-AC pre-functionalized with cell adhesion peptides (RGD). A lyophilized HA-AC aliquot (6 mg) was dissolve in 75 μL of 0.3 M TEOA buffer. Lyophilized aliquots of the RGD peptides (0.1 mg/vial) were diluted in 0.3 M TEOA (pH = 8.67), then mixed with HA-AC solution, and finally kept for reacting for 25 min at 37 °C to get the HA-RGD solution. The cells were split from the petri dish bottom by 0.25% trypsin-EDTA, and the resuspended cell solution was placed on ice. A lyophilized aliquot of the crosslinker (1.0 mg) was then diluted in 20 μL of 0.3 M TEOA buffer (pH = 8.4) immediately before mixing with HA-AC solution, HA-RGD (final concentration of 150 μM RGD), and the cell solution (3000–8000 cells per μL of final gel volume). The gel precursor solution was pipetted onto a sigmacoted glass slide in drop-wise, then clamped with another sigmacoted slide with plastic cover-slip spacers, and finally incubated for 30 min at 37 °C to allow for gelation. The final gel was swelled in culture media before being placed inside 96-well plates for the following long-term culture.

4.2.4 Rheology Measurement of Hydrogel

The storage (G') and loss modulus (G") were measured with a plate-to-plate rheometer (Physica MCR, Anton Paar, Ashland, VA) using a 8 mm plate under a constant strain of 0.01 and frequency ranging from 0.1 rad/s to 10 rad/s. Hydrogels were made as detailed above and cut to a size of 8.0 mm in diameter to fit the plate. A humid hood was used to prevent the hydrogel from drying and the temperature was kept at 25 °C.

4.2.5 Fabrication of 2D Interface

Different from the common one-dimensional (1D) interface (Figure 4.1a), the two-dimensional (2D) interface with a stereo pattern was fabricated as shown in Figure 4.1b. Here we use 2D channel in the stiff gel that has a width of 500 µm and the height is kept around 200–500 µm, and the interval distance between the adjacent channels is from 500 µm to 1000 µm. The width, interval, and height are indicated in Figure 4.1c. The cell density inside soft gel was tested ranging from 3000/µL to 7000/µL. The VMCs were uniformly distributed throughout this 1~2-mm-deep soft-gel matrix with the stimuli of microstructure interface inside the hydrogel. The steep gradient of cell concentrations around the inner interface of hydrogel was gradually formed after the cells were initially encapsulated only in the soft gel. The cells inside the soft gel partially migrated into the stiff gel, yet did not occupy much volume of the stiff gel as shown in Figure 4.1d.

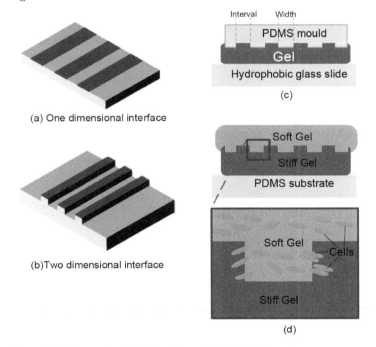

(a) One dimensional interface

(b) Two dimensional interface

(c)

(d)

Figure 4.1 Schematics of 2D interface inside 3D hydrogel.

4.2.6 3D Imaging

Using confocal laser-scanning microscope, we generated 3D images of VMC multicellular architecture, elaborating volumetric visualization of the multicellular structure in 3D space, as well as efficient quantitative analysis of the structure. For multi-cellular architecture inside the 3D composite HA hydrogel fixed between a sigmacoted glass slide and a top coverslip, the confocal laser-scanning microscope (Leica TCS SP2, Germany) was utilized to obtain the image sequences of VMC multi-cellular architecture in the vertical direction. An objective with a magnification of 10X and numerical aperture (N.A.) of 0.3 was used to primarily locate the area of interest with in the sample, while an objective (HCX PL APO CS, OIL) of 40.0X and 1.25 N.A. was used during the imaging process. The Alexa Fluor 488 Phalloidin was detected upon excitation at 488 nm (40% laser line power set in AOTF) with an emission maximum around 520 nm through a band-pass filter. The pinhole size was 0.998 airy. Hundreds of images were obtained in Z-direction via the XYZ scan mode with scanning speed at 400 Hz. The Voxel-size is 0.732 (Width [mm]) × 0.732 (Height [mm]) × 1.000 ([or 2.000] Depth [mm]). The obtained image sequences were stacked and processed by the 3D reconstruction software Amira 5.2 Resolve RT (trial version), allowing precise volumetric visualization of the multi-cellular structure in 3D space, as well as efficient quantitative analysis of the structures.

4.3 Experiment Results

4.3.1 Multicellular Network and Branching Structures inside HA Hydrogels

Vascular mesenchymal cells spread out with fingerlike projections after one day and they were persisted this morphology throughout the 7-day culture period. VMCs started remarkable interaction with neighboring cells as early as day 2, which became more pronounced between days 3 and 6. On days 7–8, the cells inside HA hydrogel were fixed by 4% PFA, and stained by fluorescent dye, which can render the F-actin inside the cells. The hydrogel preserved a microdisk-like shape until days 7–8 without any apparent degradation, and it still had a diameter of around 5 mm just as that on the day 1. The description below is a typical case of the cellular structures in a HA hydrogel we made. In the central region of the hydrogel microdisk, as shown in Figures 4.2a and b, the multicellular networks are formed, which consist of plenty of branches and sub-branches. Each branch and sub-branch is formed by locally multicellular aggregation. As shown in Figures 4.2b and c, the F-actins have significant uneven distribution in each branch (bright color indicates high density of F-actins, while dark color indicates low density of F-actins). Moreover, the uneven distributions of F-actins exist both along the branching direction (Figure 4.2b) and the radial direction of the cross-sections of individual branches indicated in Figure 4.2c. For the branching process, there seems no specific order in which the stem, branches, and sub-branches appear according to the continuous observation in the culture duration. These three parts (stem, branches, and sub-branches) seem to emerge at the same beginning time, aggregate respectively in local regions, and then connect with each other under combinational stimuli of various physical and chemical signals in 3D. Furthermore, the plane in which the second branching occurs usually has an intersection angle relative to the plane for the first branching as presented in Figure 4.2c. These intersection angles range from 10 degrees to nearly 90 degrees. The joint of the main stem and two branches as indicated in Figure 4.2c has much larger density of F-actins than that in other regions, and the cells aggregate more fiercely inside the joint. On the other hand, in the marginal region of the hydrogel microdisk, a much more aggregated multicellular structure was found as show in Figures 4.2d and e. Several cell aggregates gradually connected with each other and finally formed a large multicellular dense structure after 7–8 days, and this dense structure has an irregular and rough appearance in 3D space indicated by the red circle in Figure 4.2e. It seems to have a much larger cell density and longitudinal dimension compared to the characteristic scale of cell density and single branch cross-section of the network structure, as shown in Figures 4.2b and c. This multicellular dense structure also has branching morphology as shown in the enlarged view (Figure 4.2f), whereas geometries of their cross-sections are irregular and not similar to the round shape of the individual branch cross-sections (Figure 4.2c).

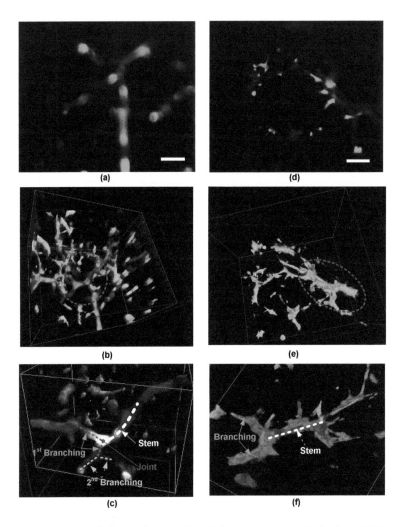

Figure 4.2 Morphology of the multicellular structure in 3D hydrogel. The cellular F-actins inside the hydrogel sample are stained by Alexa Fluor 488 Phalloidin. (a) and (c) are planar images, while (b) and (c) are reconstructed 3D images using 200 images in the normal direction of the planes shown by (a) and (c), respectively. Bright color indicates high density of F-actins, while dark color indicates low density of F-actins. Experiment condition: 3% HA, 150 μM RGD, $r = 0.6$, initial cell density = 6000/μL, day 7. The scale bar lengths in (a) and (d) are 100 μm. The dimension of 3D frame boxes in (b) and (e) is 900 μm (Length) × 700 μm (width) × 500 μm (Depth); (c) and (f) present the enlarged views through other perspective angles for the regions indicated by the red circles in (b) and (e), respectively.

4.3.2 Multicellular Network inside HA Hydrogels with Low Stiffness and Higher Stiffness

As shown in Figure 4.3, the morphologies of multicellular structures of VMCs in the uniform gels did not indicate any aggregated spheres of VMCs, neither in the homogeneous gel with lower-stiffness (Figure 4.3b) nor that with higher-stiffness (Figure 4.3c). These experimental results could serve as control group of the results that provided a foundation for the following comparison. Figure 4.3e shows that the cross-connected networks of cells were formed on the horizontal interfacial plane within region A of the composite gels, shown in Figure 4.3d. There are only small and uniformly distributed aggregates, as shown in Figures 4.3e and f. Since the proportion of HA in the total hydrogel is 2.6%, the stiffness of the soft hydrogel may be much smaller than that cell required, so lots of cells preferred to attached on the horizontal interface between the upper soft and lower stiff hydrogels, as shown in Figures 4.3e and f.

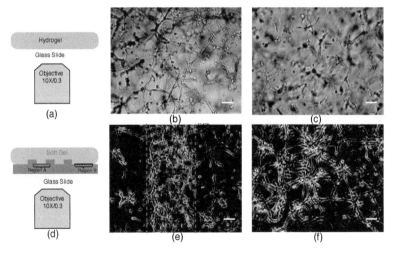

Figure 4.3 Phase contrast images using common inverted microscope (a) for VMCs in uniform soft gel (b), uniform stiff gel (c), on the horizontal interfacial plane within region A (d), in composite gels (e), and (f) shows the multicellular structures on the horizontal plane within region B (see (d)). Scale bar is 100 μm. The parameters for the soft and stiff gels: 2.6% HA, 80 μm RGD, $r = 0$, 38, and 0.7, respectively.

4.3.3 Controllable Large-Dimensional Tube Formation at Interface of High-Stiffness and Low-Stiffness Gels

The morphology of multicellular 3D structures in the composite hydrogel consisting of patterned high-stiffness and low-stiffness gels is quite different from that in uniform hydrogel stated aforementioned. In this patterned composite hydrogel, the cells were uniformly encapsulated in the low-stiffness gel at the initial time, and cells spread and highly elongated after 24 h. The cells noticeably migrated toward and began to aggregate around the interface on day 2 (Figure 4.4a), and became more pronounced on day 4, and obvious big spots with a diameter of more than 100 µm along the interface line were observed on day 8 (Figure 4.4a). At this time, the spots had already become a cylinder-like structure. The spots reached the largest dimensions on days 10–11 (Figures 4.4a–c). The cylinder-like structure seemed no significant further growing after day 10. The sample was fixed by 4% PFA on days 11–13. The most significant phenomenon in this 3D composite gel is the formation of large-dimensional upright tube around the interfaces (planner images are shown in Figures 4.4b and c). These large-dimensional orientated tubes have never been found in the uniform HA hydrogels, although large value ranges of synthesis parameters (such as concentration of HA (2–5%), RDG (50–1000 mM), and crosslinkers ($r = 0.3$–0.8)) have been tested.

Confocal laser-scanning microscopy revealed that the straight multicellular tubes are always parallel to the vertical interfacial plane inside the composite gel, and it has continuous lumen (Figures 4.4d–g). Moreover, one end of the tube is covered and sealed by a dome-shape "lid", which is a dense multicellular structure. This multicellular tube has an outer diameter larger than 160 µm, a wall thickness of 15–50 µm, and a length of more than 200 µm. The tube morphology here (Figures 4.4d–g) has noteworthy advantages compared to the self-organized capillary-like tubal structures [33, 34] reported in the past. The diameters of the tubes formed in previous capillary-like tubes seem always less than 50 µm and their size distribution and orientations cannot be actively and quantitatively controlled. In contrast, the diameter and the elongating direction of the multicellular tube in this study can be controlled by varying the patterned channel width and depth, and the direction of interfacial

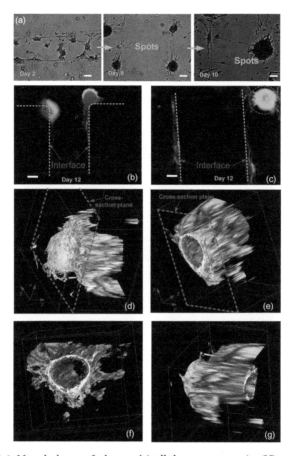

Figure 4.4 Morphology of the multicellular structure in 3D composite hydrogel consisting of patterned high-stiffness gel and low-stiffness gel (3.0% HA, 150 mM RGD, r = 0.3, 0.7, respectively); (a)–(d) are fluorescent images at the edges of the stiff gel and soft gel (culture duration can be 10–14 days). The F-actins of cells are stained by Alexa Fluor 488 Phalloidin and cellular nuclei are stained by DAPI. Scale bar length in (a)–(c) is 100 μm; (d)–(g) are 3D reconstructive images of multicellular tube structure. The dimension of 3D outermost frame box in (d) and (e) is 375 μm (length along x-axis) × 375 μm (width along y-axis) × 264 μm (depth along z-axis); (e) is the remaining part of the tube structure after the left-side part of the cross-section plane shown in (d) is removed. The cylinder is cavate inside; (f) and (g) present the views through other perspective angles for multicellular structure in (e). Reproduced from Zhu X, Gojgini S, Chen TH, et al. The 3D tubular structure self-assembled by VMCs at stiffness interfaces of hydrogels. *Biomedicine & Pharmacotherapy* 2016; 83: 1203–1211. Copyright © 2016-2021 Elsevier Masson SAS. All rights reserved.

plane, respectively, because the tube diameter seems strongly influenced by the channel width of high-stiffness gel, and the tube elongating direction is always approximately parallel to the vertical interfacial plane.

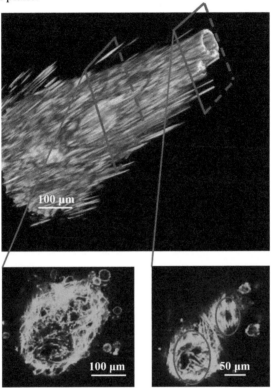

Figure 4.5 Morphology of the tubular microstructure composed of cells in 3D composite hydrogel. The tubular microstructure had a varied diameter. The F-actins inside the cells were stained and the different colors (brown and green) had been used to render the structure. The composite hydrogel consisted of patterned high-stiffness gel and low-stiffness gel.

The tubes described here actually may not be the strictly defined tube, they are the hollow architecture composed of numerous cells and they had the morphology that stretched in one direction. In some of our experiments, the tubular structure did not have a constant diameter, but had a varied diameter as shown in Figure 4.5. The different cross-sections of the hollow structure had been snapped and we can see that one tube had split into two small hollow tubes. The F-actins inside the cells were stained and the different colors (brown and green) had been used to render the structure as shown in Figure 4.5. This experiment also used the composite hydrogel described above and the composite hydrogel consisted of patterned high-stiffness gel and low-stiffness gel.

The mechanism for this large-dimensional tube formation in the composite gel could be initiated by a delicate orchestration of multiply cues, including chemical, physical, mechanical, and biological cues. In this book, we would like to use a varied model based on Turing's diffusion-reaction theory that integrating the physical, chemical, and biological cues by using mathematical language. In the next chapter, we will model this biophysical phenomenon by using a set of partial differential equations chosen to describe the diffusion and reaction of cytokines, growth factors, and cells inside a composite structure of hydrogel that incorporating structure parts with different stiffness.

4.4 Summary

Studying the cellular response at the interface of the hierarchical tissue architectures is very important for the multicellular behaviors at the interface of different types of ECM, which are crucial for recapitulating architectural features of natural tissue. Here, we created 3D self-organized multi-cellular structures mimicking 3D features of tubular tissues. It demonstrates that 3D multi-cellular tubular structures can be derived by the self-organization of a single type of stem cells (VMCs) in rationally devised composite hydrogels. It lays a foundation for regenerating injured vasculatures in organs, such as hearts, lungs, and kidneys, more efficiently and at lower cost. This self-organization methodology based on constructing the stiffness interfaces and gradients of cell concentrations in the hydrogel opens up new vistas for generating tubular tissues with an unprecedented level of rational controllability. These generated multi-cellular structures will be proved to reveal a Turing-type mechanism based on the reaction-diffusion-advection of morphogens and cells in the next chapter.

Acknowledgments

This work is supported by the National Natural Science Foundation of China (Grant No. 51505127) and Fundamental Research Funds for the Central Universities (Grant Nos. 2018B22414 and 2015B04414) of China.

References

1. Hoang, P.; Ma, Z. (2021). Biomaterial-guided stem cell organoid engineering for modeling development and diseases. *Acta Biomater.,* **132**, pp. 23–36.

2. Gjorevski, N.; Lutolf, M. P. (2017). Synthesis and characterization of well-defined hydrogel matrices and their application to intestinal stem cell and organoid culture. *Nature Protocols*, **12**, pp. 2263–2274.

3. Rizzoti, K.; Lovell-Badge, R. (2011). Regenerative medicine: Organ recital in a dish, *Nature,* **480**, pp. 44–6.

4. Bashur, C. A.; Venkataraman, L.; Ramamurthi, A. (2012). Tissue engineering and regenerative strategies to replicate biocomplexity of vascular elastic matrix assembly, *Tissue Engineering Part B: Reviews,* **18**, pp. 203–217.

5. Huh, D.; Matthews, B. D.; Mammoto, A.; Montoya-Zavala, M.; Hsin, H. Y.; Ingber, D. E. (2010). Reconstituting organ-level lung functions on a chip, *Science,* **328**, pp. 1662–8.

6. Berthiaume, F.; Maguire, T. J.; Yarmush, M. L. (2011). Tissue engineering and regenerative medicine: History, progress, and challenges, *Annu Rev Chem Biomol Eng,* **2**, pp. 403–430.

7. Haycock, J. W. (2011). 3D cell culture: A review of current approaches and techniques, *Methods Mol Biol,* **695**, pp. 1–15.

8. Tsang, V. L.; Bhatia, S. N. (2004). Three-dimensional tissue fabrication, *Adv Drug Deliv Rev,* **56**, pp. 1635–47.

9. Shimizu, T.; Yamato, M.; Kikuchi, A.; Okano, T. (2003). Cell sheet engineering for myocardial tissue reconstruction, *Biomaterials,* **24**, pp. 2309–16.

10. Yang, J.; Yamato, M.; Shimizu, T.; Sekine, H.; Ohashi, K.; Kanzaki, M.; Ohki, T.; Nishida, K.; Okano, T. (2007). Reconstruction of functional tissues with cell sheet engineering, *Biomaterials,* **28**, pp. 5033–43.

11. Odde, D. J.; Renn, M. J. (2000) Laser-guided direct writing of living cells. *Biotechnol Bioeng,* **67,** pp. 312–8.

12. Mironov, V.; Boland, T.; Trusk, T.; Forgacs, G.; Markwald, R. R. (2003). Organ printing: Computer-aided jet-based 3D tissue engineering, *Trends Biotechnol,* **21**, pp. 157–61.

13. Ho, C.-T.; Lin, R.-Z.; Chang, W.-Y.; Chang, H.-Y.; Liu, C.-H. (2006). Rapid heterogeneous liver-cell on-chip patterning via the enhanced field-induced dielectrophoresis trap, *Lab on a Chip,* **6**, pp. 724–734.

14. Jakab, K.; Norotte, C.; Marga, F.; Murphy, K.; Vunjak-Novakovic, G.; Forgacs, G. (2010). Tissue engineering by self-assembly and bio-printing of living cells, *Biofabrication,* **2**, pp. 022001.

15. Liu, J. S.; Gartner, Z. J. (2012). Directing the assembly of spatially organized multicomponent tissues from the bottom up, *Trends Cell Biol.,* **22**, pp. 683–91.

16. Nelson, C. M.; Bissell, M. J. (2005). Modeling dynamic reciprocity: Engineering three-dimensional culture models of breast architecture, function, and neoplastic transformation, *Semin. Cancer Biol.,* **15**, pp. 342–52.

17. Benjamin, M.; Evans, E. J.; Copp, L. (1986). The histology of tendon attachments to bone in man, *J. Anat.,* **149**, pp. 89–100.

18. Lu, H. H.; Jiang, J. (2006). Interface tissue engineering and the formulation of multiple-tissue systems, *Adv. Biochem. Eng. Biotechnol.,* **102**, pp. 91–111.

19. Lee, J.; Cuddihy, M. J.; Kotov, N. A. (2008). Three-dimensional cell culture matrices: State of the art, *Tissue Engineering Part B-Reviews,* **14**, pp. 61–86.

20. Trkov, S.; Eng, G.; Di Liddo, R.; Parnigotto, P. P.; Vunjak-Novakovic, G. (2010). Micropatterned three-dimensional hydrogel system to study human endothelial-mesenchymal stem cell interactions, *J. Tissue Eng. Regen. Med.,* **4**, pp. 205–15.

21. Jakab, K.; Norotte, C.; Damon, B.; Marga, F.; Neagu, A.; Besch-Williford, C. L.; Kachurin, A.; Church, K. H.; Park, H.; Mironov, V.; Markwald, R.; Vunjak-Novakovic, G.; Forgacs, G. (2008). Tissue engineering by self-assembly of cells printed into topologically defined structures, *Tissue Eng Part A,* **14**, pp. 413–421.

22. Kehr, N. S.; Riehemann, K. (2016). Controlled cell growth and cell migration in periodic mesoporous organosilica/alginate nanocomposite hydrogels, *Advanced Healthcare Materials,* **5**, pp. 193–197.

23. Simona, B. R.; Hirt, L.; Demko, L.; Zambelli, T.; Voros, J.; Ehrbar, M.; Milleret, V. (2015). Density gradients at hydrogel interfaces for enhanced cell penetration, *Biomaterials Science, 3*, pp. 586–591.

24. Kim, T. H.; An, D. B.; Oh, S. H.; Kang, M. K.; Song, H. H.; Lee, J. H. (2015). Creating stiffness gradient polyvinyl alcohol hydrogel using a simple gradual freezing-thawing method to investigate stem cell differentiation behaviors, *Biomaterials, 40*, pp. 51–60.

25. Rao, S. S.; Bentil, S.; DeJesus, J.; Larison, J.; Hissong, A.; Dupaix, R.; Sarkar, A.; Winter, J. O. (2012). Inherent interfacial mechanical gradients in 3D hydrogels influence tumor cell behaviors, *PLoS One, 7*, pp. e35852.

26. Snyder, T. N.; Madhavan, K.; Intrator, M.; Dregalla, R. C.; Park, D. (2014). A fibrin/hyaluronic acid hydrogel for the delivery of mesenchymal stem cells and potential for articular cartilage repair, *Journal of Biological Engineering, 8*, 10:1–11.

27. Zhu, X.; Gojgini, S.; Chen, T. H.; Fei, P.; Dong, S.; Ho, C.-M.; Segura, T. (2017). Directing three-dimensional multicellular morphogenesis by self-organization of vascular mesenchymal cells in hyaluronic acid hydrogels, *Journal of Biological Engineering, 11*, pp. 12.

28. Lei, Y.; Gojgini, S.; Lam, J.; Segura, T. (2011). The spreading, migration and proliferation of mouse mesenchymal stem cells cultured inside hyaluronic acid hydrogels, *Biomaterials, 32*, pp. 39–47.

29. Chen, T. H.; Hsu, J. J.; Zhao, X.; Guo, C. Y.; Wong, M. N.; Huang, Y.; Li, Z. W.; Garfinkel, A.; Ho, C. M.; Tintut, Y.; Demer, L. L. (2012). Left-right symmetry breaking in tissue morphogenesis via cytoskeletal mechanics, *Circulation Research, 110*, pp. 551–559.

30. Chen, T. H.; Zhu, X.; Pan, L.; Zeng, X.; Garfinkel, A.; Tintut, Y.; Demer, L. L.; Zhao, X.; Ho, C. M. (2012). Directing tissue morphogenesis via self-assembly of vascular mesenchymal cells, *Biomaterials, 33*, pp. 9019–26.

31. Zhu, X.; Ding, X. (2017). Study on a 3D hydrogel-based culture model for characterizing growth of fibroblasts under viral infection and drug treatment, *SLAS Discovery, 22*, 626–634.

32. Zhu, X.; Gojgini, S.; Chen, T. H.; Teng, F.; Fei, P.; Dong, S.; Segura, T.; Ho, C.-M. (2016). Three dimensional tubular structure self-assembled by vascular mesenchymal cells at stiffness interfaces of hydrogels, *Biomedicine & Pharmacotherapy, 83*, pp. 1203–1211.

33. Sudo, R.; Chung, S.; Zervantonakis, I. K.; Vickerman, V.; Toshimitsu, Y.; Griffith, L. G.; Kamm, R. D. (2009). Transport-mediated angiogenesis in 3D epithelial coculture, *FASEB J.,* **23**, pp. 2155–2164.

34. Anderson, S. M.; Siegman, S. N.; Segura, T. (2011). The effect of vascular endothelial growth factor (VEGF) presentation within fibrin matrices on endothelial cell branching, *Biomaterials,* **32**, pp. 7432–43.

Chapter 5

Modeling Cellular Self-Assembly at Patterned Interfaces inside Hydrogel via Turing's Reaction-Diffusion Frame

5.1 Introduction

Self-organization paradigm in tissue morphogenesis is emerging its potential power [1–5] for regenerating injured tissue for regenerative medicine. It overcomes some limitations associated with the usage of scaffolds with specific structures since the fabricated mechanical scaffolds cannot totally possess the complexity of native tissues. It is also being realized that the better approach for repair or regenerating tissues might need to rely on the self-organizing properties of cells/ tissues and the innate regenerative capability of the organism itself [6, 7], because the self-organizing formation of multicellular structures in vitro probably could potentially present biological processes similar to those that occur in vivo, which would help lead to the recapitulation of morphological structure and organization of native tissues [8–11].

Tubular or hollow structures serve as fundamental units in most organs. For example, the lung, kidney, and vasculature are composed primarily of tubes [6], such as trachea, arteries, and lymph vessels. Also, the small cavities exist in many inner structures of the organs. Therefore, artificially constructing hollow structures including tubes are urgently needed for advancing regenerative medicine. Recently,

Self-Organized 3D Tissue Patterns: Fundamentals, Design, and Experiments
Xiaolu Zhu and Zheng Wang
Copyright © 2022 Jenny Stanford Publishing Pte. Ltd.
ISBN 978-981-4877-77-0 (Hardcover), 978-1-003-18039-5 (eBook)
www.jennystanford.com

researchers have used a stress-induced rolling membrane technique to fabricate tubular structures with configurable sizes [7]. This method can obtain microtubes with multilayers consisting of different types of cells. However, the geometry features, such as diameters and wall thicknesses, are limited by the membrane thickness and number of rolling layers, creating obstacles to constructing smaller and thinner-wall tubular structure. A tubular structure can also be formed with a monolayer of cells along the fiber direction in a human umbilical vein endothelial cell (HUVEC) collagen microfiber-shaped construct [8]. This is a typical scaffold-based approach yet with a deformable fiber as the scaffold. Moreover, 3D printing can produce networks of carbohydrate glass serving as cytocompatible sacrificial scaffold to construct vascular architectures [9]. The hollow cellular structures can also be constructed by 3D bioprinter using the biomimetic components made of biomaterials such as hydrogels encapsulating cells [3, 10]. Nevertheless, using structured scaffolds or hydrogel to "mold" 3D tubular structure or networks still pose concerns considering the biocompatibility for applied materials or the structural complexity achievable. Moreover, artificially-predefined patterns at initial times are usually disorganized by cellular self-organization, such as cellular alignment or migration, in the subsequent tissue development, creating uncertainly for those artificial attempts.

Overall, these endeavors in tissue regeneration are trying to bridge structures of histological architecture or gross anatomy with the corresponding function of cell biology and physiology. Building the cell-hydrogel constructs or assembling of these hybrid constructs have provided flexible approaches for directing the cellular organization. However, this paramount challenge of rationally engineering tissues/organs that substitute the native tissue/organs is not fully conquered. One of the reasons is the organizing behaviors of cells such as cell migration, proliferation, and differentiation may still somewhat disorganize and frustrate the artificial attempts of shaping the engineered extracellular matrix (ECM) materials, such as hydrogels, or guiding the distribution of cells after the repopulation of cells inside the printed hydrogel matrix. Another reason could be explained as follows: Engineering of cell-hydrogel constructs mainly relies on the extra forces, action, or special control. People pre-designed the macro geometrical structures based on empirical protocols, and then engineered the 3D constructs encapsulating cells with specific geometry or distributions to expect the desired tissue or organoids.

Yet, the emergence of the desired structural morphologies is not easy to be observed without enough times of trials, because the final morphologies in complex biological systems cannot be predicted deterministically and rationally, due to insufficient understanding of the mechanism. Thus, it needs many experiments and tests, which is essentially a trial-and-error method. Consequently, it seems more attractive to utilize a predictive theory framework to guide the spontaneous cellular self-organizing process in a desired manner. It could potentially become another innovatory solution for tissue/organoid regeneration [2, 3].

In order to theoretically study or experimentally control cells in the 3D interactions [12] or 3D self-organization [13], several types of mathematical models [14–17] based on Turing's reaction-diffusion (RD) mechanism [18, 19] were created to explore the mechanisms of the formation of multicellular structures or tissues and predict the possibly organized structures. This theory was first proposed by Alan Turing in 1952 [18] , which hypothesized that the self-organization process of the biological patterns involves two types of morphogens: activator and inhibitor. The RD model commonly requires a morphogen pair [20] (an activator-inhibitor pair), in which the inhibitor diffuses more rapidly than its activator. Based on the simplified molecular diffusion-reaction mechanism, the final concentration distribution of activators and inhibitors can be computed and predicted, and then the spatial distribution of cell density can be obtained accordingly. For example, the area with high concentration of activators also has high cell density, which makes it possible to approximately compute the morphological patterns of some tissues. Turing instability in the RD system can induce spatial patterns, such as spots, stripes, hole patterns, and more complex 3D patterns, which is applied to model the spatial patterns in natural animals and plants [21, 22] or the tissue morphogenesis [23–25]. When utilizing a Turing instability-based framework to investigate the hollow tubular multicellular structure formation, the 3D material substrate should be elaborately designed in the physical and geometry perspectives, because the multicellular or ECM architectures inside the tissue usually exhibit features of gradient distribution. Spatial (structural) and temporal gradients regulate various cell behaviors, such as proliferation, migration, and differentiation, during development, wound healing, and cancer [26]. The structural or mechanical gradients also influence the function of ubiquitous tissue-to-tissue interfaces, such as mediating load transfer between two distinct types of tissue, sustaining the heterotypic cellular communications and homeostasis [27–29]. Therefore, in order to enhance the efficacy of the theoretical framework based on Turing instability, the cellular self-assembly behavior under certain concentration gradients of cells and bio-signals, or around the interface of different extracellular matrices should be studied.

Here, we theoretically study a simulation model that describes an engineered composite hydrogel structure with interfacial planes inside it to serve as the 3D culture matrix to sculpt the 3D multicellular structures. We use vascular mesenchymal cells (VMCs), as the research objective mainly because VMCs have recognized activator (Bone Morphogenic Protein-2 [BMP-2]) and its inhibitor (Matrix Gla Protein [MGP]) in a RD system [18, 19]. Specifically, we computationally investigate the 3D pattern formation process of VMCs at the interfaces between hydrogels with different components and mechanical properties. In this dynamic system, the VMCs interact with BMP-2 and MGP, which serve as a pair of activator and inhibitor. Based on the biochemical interactions of BMP-2, MGP, and cells, a RD mathematical model is derived in 3D space. Accordingly, the simulation model is computed in a 3D domain having the same geometrical feature as the 3D hybrid hydrogel consisting of lower-stiffness gel and higher-stiffness gel.

5.2 Theoretical Model

Self-organization of cells usually can be achieved in hydrogel system [1–3]. Here, VMCs serve as a typical study object. VMCs can secrete two types of morphogens (proteins), slowly-diffusing activators (such as BMP-2 and BMP-4), and rapidly-diffusing inhibitor (such as MGP) [30, 31]. Based on our previous study [20], we modeled the cell-hydrogel system as the reaction (using Gierer and Meinhardt kinetics) and diffusion of the autocatalytic and slowly-diffusing equivalent activator, U, its rapidly-diffusing equivalent inhibitor, V, and cell density, n, reflecting proliferation, cytokinetic motility, and chemotactic migration with respect to activator U, as functions of a 3D domain (x, y, z) in dimensionless forms:

$$\frac{\partial U}{\partial t} = D\nabla^2 U + \gamma\left[\frac{pnU^2}{V\left(1+mU^2\right)} - cU\right] \tag{5.1}$$

$$\frac{\partial V}{\partial t} = \nabla^2 V + \gamma\left[bnU^2 - eV\right] \tag{5.2}$$

$$\frac{\partial n}{\partial t} = q\nabla^2 n - \chi\left[\nabla\cdot\left(\frac{n}{\left(k_n+U\right)^2}\nabla U\right)\right] + r_n n\left(1-n\right) \tag{5.3}$$

In Eqs. (5.1)–(5.3), U, V, and n are dimensionless concentrations of activator (U), inhibitor (V), and cells (n) as functions of space coordinate (x, y, z) and time (t); c and e are the degradation rate of activator (such as BMP-2) and inhibitor (such as MGP), and b is the coefficient representing the relative production of inhibitor to activator; D and q are dimensionless diffusion coefficients of morphogens and cells, and they are the ratios of diffusion coefficients for activator-to-inhibitor, and cells-to-inhibitor, respectively; γ is a scaling factor related to domain size, biosynthetic timescale, and inhibitor diffusivity; m and k_n are the dimensionless constants for the saturation of autocatalytic production of activator and chemotaxis respectively; χ is the dimensionless chemotactic coefficient; r_n is the dimensionless form of maximum of cell proliferation rate.

Studying the morphogenesis of the multicellular system with heterotypic cells is essential for the understanding of the mechanism underlying this coherent pattern formation, which could eventually lead us to control the pattern evolution. In this study, we are trying to explain the mechanism underlying the self-assembly of the VMCs around the interface in 3D hydrogel by using a modified mathematical model of two-dimensional (2D) diffusion-reaction equations [18]. In previous work [32], the mechanism underlying the above morphogenesis of VMCs has been modeled with the above modified diffuse-reaction equations.

Here, the above RD model is applied to mathematically describe a 3D hydrogel system, as shown in Figure 5.1a. The 3D hydrogel consists of two gels having distinct mechanical stiffness, which are indicated as lower-stiffness gel and higher-stiffness gel. These two parts of gels are matched tightly with each other to form physical interfacial planes between them. In this study, the cells and the morphogens secreted by them only exist in the region of lower-stiffness gel at the initial time, as shown in Figure 5.1a, aiming to achieve a striking gradient of cell density around the interfacial planes. This hybrid hydrogel system is used to approximately mimic the ubiquitous situation, and the gradients of structural and mechanical properties in native tissues. The mathematical model has corresponding parameters to describe the diffusion and reaction process of the activators, inhibitors, and cells in this hybrid hydrogel. As shown in Figure 5.1b, the cell migrations are divided into two types: one is the migration toward the location with high-cell density (cell cluster); the other is the migration away from the cell cluster. Each of these two types of cellular migration is caused by the combined effect of cell diffusion and chemotaxis. For example, the cell migration toward the cluster is not only due to the cell chemotaxis, but also influenced by cell diffusion. The migration direction is finally toward the cellular cluster because the cell chemotactic migration played a dominant role, which means the magnitude of chemotactic term $\left(-\chi\left[\nabla \cdot\left(\left(n_1/(k_n+U)^2\right)\nabla U\right)\right]\right)$ is larger than diffusion term $(q\nabla^2 n)$ in Eq. (5.3). The simulation based on the mathematical model (Eqs. (5.1–5.3)) was performed in a 3D computing domain, possessing the same geometrical feature as the 3D hydrogel shown in Figure 5.1a.

In the simulation model, the dimensionless diffusion coefficients of morphogens (D) in higher-stiffness gel and lower-stiffness gel are denoted as D_H and D_L, respectively; the dimensionless diffusion coefficients of cells (q) in higher-stiffness gel and lower-stiffness gel are denoted as q_H and q_L, respectively; the dimensionless chemotactic coefficient of cells (χ) in higher-stiffness gel and lower-stiffness gel are denoted as χ_H and χ_L, respectively; The parameter evaluations are achieved by setting $D_H = 0.005$, $q_H = 5\times10^{-5}$, $\chi_H = 5\times10^{-5}$; $D_L = 0.008$, $q_L = 1\times10^{-4}$, and $\chi_L = 1\times10^{-4}$. As shown in Figure 5.1c, at the initial stage of the simulative evolution, the overwhelming majority of cells stayed in the region of lower-stiffness gel, and the blue regions have no cells. A sharp density gradient of cells at the small interfacial regions started to form. There were no patterns of cells at the initial stage.

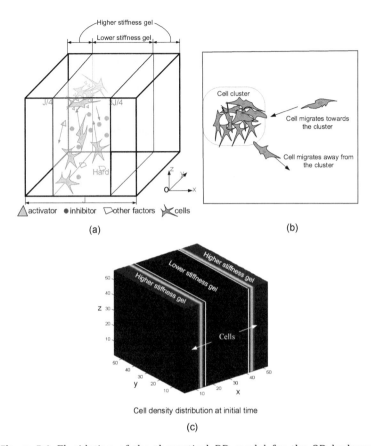

Cell density distribution at initial time

(c)

Figure 5.1 Elucidation of the theoretical RD model for the 3D hydrogel consisting of two gels having distinct mechanical stiffness, matched tightly with each other to form a physical interfacial surface between them. (a) Distribution of activator, inhibitor, and cells at initial time. (b) The illustration of cellular migrations toward or away from the cellular aggregation in the 3D hydrogel. (c) Distribution of cell density in 3D computational domain at the initial stage of the simulative evolution. Concentration gradient of cells emerges around the interfacial plane of lower-stiffness gel and higher-stiffness gel. This figure is reproduced with permission from [33].

5.3 Simulation Results and Discussion

The formation of cylinder tubes and hollow spheres by self-organization of cells in the composite hydrogel were simulated and shown in Figure 5.2. At preliminary stage, the cells virtually distributed in the gel domain with lower-stiffness only, and gradually formed a cell density gradient across the interfacial plane, as shown in Figure 5.2a. At the simulation time instant of $t = 1$, the cell aggregated into the perforated sheet structure around the interfacial surface. When $t = 1.5$, the multicellular structure evolved into more locally aggregated bough-like structure with hollow inner cave, as shown in Figure 5.2c. In the following evolution, the connected bough-like tubes gradually lost the connections among them, as shown in Figure 5.2d, and a stand-alone vertical tube emerged as shown in Figure 5.2e. During the evolution of self-formed multicellular structure, the alteration of cellular density distribution, due to the chemotaxis of cells was significant slower than the change of concentration distribution of activators (molecules), so we employed a low value of chemotactic coefficient ($\chi = 10^{-4}$) in the simulation. The diffusion effect of cells might exceed the chemotactic effect of cells, which indicated the cells could be migrated toward the regions with low-concentration of activator (periphery sites of the vertical cylinder or spheres) during a specific phase of the evolution. Simultaneously, the sites attracting more cells had more activators produced, which, in turn, enhanced the cellular chemotactic effect at local sites, and gradually facilitated the cellular chemotactic effect and the cells' diffusion effect to reach the dynamic equilibrium state. Accordingly, the cells finally stayed on the periphery region of the tubes or spheres (i.e. tube walls or spherical shells). The axis of the cylindrical tube was approximately perpendicular to the horizontal surface in the simulation at $t = 4$, as shown in Figure 5.2f. The hollow tubes and spheres still kept their morphology after the simulation time was increased to $t = 5$, as presented in Figures 5.2g and h, which means these results are robust and can sustain for a long time in this dynamic system.

The simulation results indicate that the self-organized tubular and spherical structures could be generated in the suitably-tuned RD system based on a Turing-type mechanism. In our proposed simulation model, the emergence of self-organized tubular and spherical structures is mainly because of the matrix-to-matrix interface leading to a local variation of physical and chemical cues. Therefore, the underlying implication not only includes RD of morphogens, but also the self-formed local concentration gradients of morphogens and cells, due to the change of mechanical and biophysical properties across the matrix-to-matrix interfacial planes.

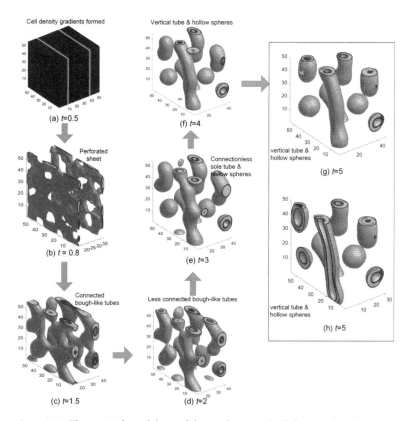

Figure 5.2 Theoretical modeling of the evolution of cell density distribution for forming the 3D self-organized tubular and spherical topology. The rendered color map indicates the distribution of cell density. Red color indicates high value of cell density. The lowest values were made transparent for clarity. (a) When $t = 0.5$, the cells distributed in the gel domain with low stiffness only. (b) At the simulation time instant of $t = 1$, the cell aggregated into the perforated sheet structure around the interfacial surface. (c) When $t = 1.5$, the perforated sheet structure evolved to connected bough-like tubes. (d) When $t = 2$, bough-like tubes had less inter-connections. (e) When $t = 3$, bough-like tubes lost the connections among them and a stand-alone vertical tube emerges. (f) When $t = 4$, the tube became more straight and the hollow spheres became more obvious. (g) Vertical tubes and hollow spheres sustained their morphologies at $t = 5$. (h) Cutaway view of the tubes and spheres self-organized by a large number of cells. The parameters for the higher-stiffness gel are assumed $D_H = 0.005$, $q_H = 5 \times 10^{-5}$, and $\chi_H = 5 \times 10^{-5}$; the parameters for the lower-stiffness gel are assumed $D_L = 0.008$, $q_L = 1 \times 10^{-4}$, and $\chi_L = 1 \times 10^{-4}$. Other parameters are estimated as $p = 0.7$, $b = 1.0$, $c = 0.04$, $e = 0.02$, $m = 0.22$, $k_n = 1.0$, and $r_n = 5.6$.

This simulation model may also be helpful for interpreting the geometry variation of tubular or spherical structure formed around the interfaces of hydrogels when the parameter is changed, so it may have the potential to guide the fabrication of tubular or sphere structures with more options for sculpting diverse typologies and dimensions.

Hollow tubes and spheres assembled by numerous cells can still emerge after the parameters of the cell-hydrogel system have been changed within moderate ranges in our simulation model. It follows that the emerged hollow structures have good robustness, thus the effectiveness of the simulation could be demonstrated. For the following simulation results, some key parameter values for the higher-stiffness gel were changed to $q_H = 7\times10^{-5}$, $\chi_H = 7\times10^{-5}$; some key parameter values for the lower-stiff gel were changed to $q_L = 2\times10^{-4}$, $\chi_L = 3.1\times10^{-4}$. These values were increased compared to the aforementioned simulation.

At preliminary stage, the cells had the almost uniform distribution, as shown in Figure 5.3a, and this scenario almost kept the same as aforementioned simulation because the cells were always supposed to settle in the gel domain with low stiffness only, as an initial spatial stimulus. This simulation had the similar results during the whole process of the reaction-diffusing dominated self-assembly of cells, although it had some different morphologies compared to aforementioned simulation at several stages, such as $t = 0.5$, 5 in Figures 5.3b–d, yet the structural geometry features were always indicating the formation process of hollow tubular and spherical structures. At the simulation time instant of $t=2.5$, the cells aggregated into the connected bough-like structures around the interfacial surface (Figures 5.3c and d). The formation of these structures could be elucidated as the gradually holing and piercing process, due to the spatial aggregation of cells. It was the pre-formation of tubular structures, and the connected bough-like structures disconnected and shrunk into disconnected bough and sphere structures ($t = 5$), and the obstructed inner lumen gradually became barrier-free and had continuous inner walls at a later time ($t = 25$), as shown in Figure 5.3f. The formed stand-alone tube along vertical direction is shown in Figures 5.3f and g. This implies more information for the generated morphology of the multicellular structure self-organized via RD process. For instance, the radial direction of tubes seems to be consistent with the direction of cell concentration gradient. As the time increased, the walls of hollow tubes evolved into denser multicellular structure and the tube axis became more straight at t = 35 compared to the previous serpentine shape (t = 5, Figure 5.3e). The tube walls presented orange color, indicating the walls had higher cell density than that in marginal regions. The spherical structures seemed to shrink and became oblate hemispheres at t = 25 (Figure 5.3f) compared to that at t = 5 (Figure 5.3e). After the comparison between the results in aforementioned figures, it follows that the hollow tube can also emerge when the parameter values change in the simulation model, such as the increase of $q_H, \chi_L, q_L,$ and χ_L. The time instants at which hollow tubes formed are different between the situations, as shown in Figure 5.2 and Figure 5.3. The lumen feature inside the multicellular structures is potentially exhibited as early as t = 2.5 as shown in Figure 5.3, with the increased value of $q_H, \chi_H, q_L,$ and χ_L. The common points between the cases in Figure 5.2 and Figure 5.3 are: (i) the cells diffused out from the low-stiffness regions and aggregated around the interfacial surfaces; (ii) the spatial aggregation of cells further resulted in holing and piercing process inside the aggregated homogeneous multicellular structure; (iii) the tubular structure emerged during the holing and piercing process, which gave rise to more disconnections between the topological features.

These simulation results could further lead to conceiving a technique to create hollow tubular and spherical multicellular structures via self-organization of VMCs. The creation of hollow tube using VMCs' self-organization provides a potential methodology to meet the urgent need for small-diameter vascular graft. The use of VMCs could bypass the limitation of the use of embryonic stem cells due to ethical concerns. These VMCs could be isolated from a patient's own tissue, thus more stable biophysical and biocompatible properties of this vascular graft could be achieved. This completely self-organized tubular structure would not involve any physical scaffold with tubular topology, so it avoids most of the existing problems in the course of scaffold seeding and degradation, such as eliciting immunogenic reaction and geometrical influence of the scaffold itself. Our study also demonstrates that the biological context at matrix-to-matrix interfaces can have other important functions of facilitating the formation of tissue structure with tubular and spherical features.

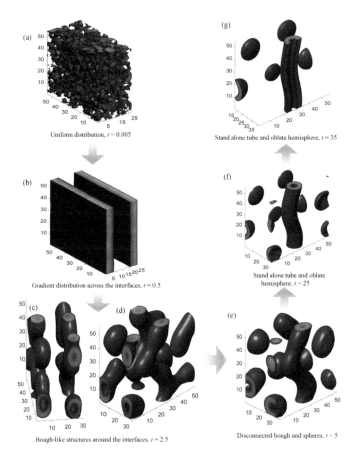

Figure 5.3 Theoretical modeling of the evolution of cell density distribution for forming the 3D self-organized tubular topology similar to the literature [34]. (a) When $t = 0.005$, the cells virtually distribute in the gel domain with low stiffness only. (b) When $t = 0.5$, most of the cells diffused and aggregated around the interfacial surfaces. (c and d) At the simulation time instant of $t = 2.5$, the cells aggregated into the connected bough-like structures around the interfacial surface. (e) When $t = 5$, the connected bough-like tubes gradually lose the connections among them. (f) A stand-alone vertical tube formed at $t = 25$. (g) The cutaway view of the tube after partial volume of it was removed. The rendered color map indicates the distribution of cell density. Orange color indicates high value of cell density, while blue color indicates low value of cell density. The lowest values were made transparent for clarity. The parameters for the higher-stiffness gel are assumed $D_H = 0.005$, $q_H = 7 \times 10^{-5}$, and $\chi_H = 7 \times 10^{-5}$; the parameters for the lower-stiff gel are assumed $D_L = 0.008$, $q_L = 2 \times 10^{-4}$, and $\chi_L = 3.1 \times 10^{-4}$. Other parameters are estimated as $p = 1.0$, $b = 1.0$, $c = 0.04$, $e = 0.02$, $m = 0.25$, $k_n = 1.0$, and $r_n = 5.6$.

It has the potential to have the cell-encapsulated hydrogels integrated into the native tissue surface to construct the key interfacial surfaces between hydrogel and native tissues, in order to the tubular tissue formation in vivo. It would have great perspective for regenerating the tubular structures, such as blood vessels, lymph vessels, trachea, or intestines, in a rationally controllable self-organizing manner.

We performed more simulations by changing some parameter values to $q_H = 7 \times 10^{-5}$, $\chi_H = 7 \times 10^{-5}$; $q_L = 2 \times 10^{-4}$, $\chi_L = 3.7 \times 10^{-4}$, and $m = 0.19$. The results are shown in Figure 5.4. The minimum time interval for computation is set 5×10^{-5}. At preliminary stage, the cells virtually distributed in the gel domain with lower-stiffness only, and gradually formed a cell density gradient across the interfacial plane, as shown in Figures 5.4a and b. When $t = 2.5$, twisted canes-like structures emerged as shown in Figure 5.4c. The "twisted canes" had hollow geometrical features, but they did not have obvious branching morphology. When $t = 5$, the thick canes-like structures had less connections between them and the lumen inside the tube became more obvious as presented in Figure 5.4d. With the increase of the simulation time, the short-length tubes became hollow spheres at $t = 25$, as shown in Figure 5.4e. The hollow spheres sustained their morphologies at $t = 35$, as shown in Figure 5.4f. As to the simulation results in Figure 5.4, the values of q and χ for both the higher-stiffness gel and lower-stiffness gel were raised, which indicated increased cell motility and proliferation rate. Cells lived in a more favorable extracellular environment further activating biosynthetic process of the cells. Therefore, t_c, reflecting the timescale of biosynthetic kinetics was assumed as a decreased value of 2600 s. r_n was estimated as 0.02 h^{-1}. D_h was estimated as an increased value of 1.2×10^{-7} cm^2/s.

Here, the dimensionless units of spatial length (L) is set $L = 1$ mm (thus the simulated cube is 1 mm $\times$ 1 mm $\times$ 1 mm), which leads to $\gamma = 393$, $r_n = 5.67$ approximately in dimensionless form. The time in real world T can be calculated as $T = (L^2/D_h)t$, which indicates the $t = 0.05, 0.5, 2.5, 5, 25,$ and 35, leads to $T \approx 1.2$ h, 11.6 h, 57.9 h, 4.8 days, 24 days, and 33.8 days, respectively. These time values are only estimated with an approximate timescale. The relative longer duration of 24 days or 33.8 days derived from this mathematical model may be utilized to interpret the development of certain tissue structure formation in vivo or the implanted tissue grafts.

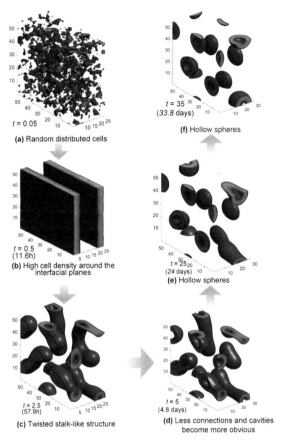

Figure 5.4 Simulation of the evolution of cell density distribution for forming the 3D self-organized short-length tubular and spherical topology [33]. The tubes and spheres are self-organized by a huge number of cells. The rendered color map indicates the distribution of cell density. Red/orange color indicates high value of cell density. The lowest values were made transparent for clarity. (a) At the simulation time instant of t = 0.5, the cell aggregated around the interfacial planes. (b) When t = 2.5, the twisted canes-like structures emerged. (c) When t = 5, the canes-like structures had less connections between them and had obvious lumens or cavities inside the aggregates. (d) The short-length tubes became hollow spheres at t = 25. (e) The hollow spheres sustained their morphologies at t = 35. The parameters for the higher-stiffness gel are assumed D_H = 0.005, $q_H = 7 \times 10^{-5}$, and $\chi_H = 7 \times 10^{-5}$; the parameters for the lower-stiff gel are assumed D_L = 0.008, $q_L = 2 \times 10^{-4}$, and $\chi_L = 3.7 \times 10^{-4}$. Other parameters are estimated as b = 1.0, c = 0.04, e = 0.02, k_n = 1.0, γ = 393, and r_n = 5.67. The symbol t denotes the time in simulation model, while the symbol T denotes the estimated time in real world.

The culture time needed for in vitro experiments may be less and within 20 days. The mathematical prediction for the time could provide the tendency and guidance for conducting in vitro experiment although it may be not accurate in the specific values.

It is followed that the short-length tubes and hollow spherical structures emerged at different times. The short-length tubes emerged much earlier than hollow spherical structures in the simulation as shown in Figure 5.4. This mathematical model has the potential to rationally elucidate that the hollow tubes with short length could emerge around the interface of hydrogels with different stiffness, which were preliminarily demonstrated by the recent experiment on 3D culture of VMCs in modified hyaluronic acid hydrogels [24]. But experimentally constructing the distributed hollow spheres composed of VMCs may need much longer time, possibly encountering difficulties in ordinary biological labs. Nevertheless, it is still possible to shorten the evolution time by optimizing the cell culture protocol and tuning the chemical and physical properties of hydrogel based on the RD framework.

5.4 Summary

The simulation model based on Turing-type mechanism is built for simulating cellular self-organization in 3D hydrogels with inner interfaces. The simulation results show the self-organized tubular and spherical structures with inner cavities can be generated around the interfacial planes inside the 3D matrices. This simulation model mathematically describes suitably-tuned RD system with the recognized activator (BMP-2) and its inhibitor (MGP) for VMCs. The generated hollow structures self-organized by cells are robust and can sustain for a long time in this dynamic simulation system. The multicellular morphology changes from the branched tubes to the short-length tubes, with the variation of parameter values. The short-length tubes also can evolve to hollow sphere-like structures after long-time cellular self-organization. Our simulations can rationally predict 3D multicellular tubular and spherical structure formation in a heterogeneous matrix, which lays a foundation for regenerating injured vasculatures or hollow structure unit in organs, such as hearts, lungs, and kidneys, in a more rational and efficient way.

Acknowledgments

This work is supported by the National Natural Science Foundation of China (Grant Nos. 51875170 and 51505127) and Fundamental Research Funds for the Central Universities (Grant Nos. B200202225, 2018B22414 and 2015B04414) of China.

References

1. Suga, H.; Kadoshima, T.; Minaguchi, M.; Ohgushi, M.; Soen, M.; Nakano, T.; Takata, N.; Wataya, T.; Muguruma, K.; Miyoshi, H.; Yonemura, S.; Oiso, Y.; Sasai, Y. (2011). Self-formation of functional adenohypophysis in three-dimensional culture, *Nature*, **480**, pp. 57–62.

2. Takebe, T.; Sekine, K.; Enomura, M.; Koike, H.; Kimura, M.; Ogaeri, T.; Zhang, R. R.; Ueno, Y.; Zheng, Y. W.; Koike, N.; Aoyama, S.; Adachi, Y.; Taniguchi, H. (2013). Vascularized and functional human liver from an iPSC-derived organ bud transplant, *Nature* **499**, pp. 481–484.

3. Athanasiou, K. A.; Eswaramoorthy, R.; Hadidi, P.; Hu, J. C. (2013). Self-organization and the self-assembling process in tissue engineering, *Annu. Rev. Biomed. Eng.*, **15**, pp. 115–136.

4. Hagiwara, M.; Peng, F.; Ho, C. M. (2015). In vitro reconstruction of branched tubular structures from lung epithelial cells in high cell concentration gradient environment, *Scientific Reports*, **5**, pp. 80541–80547.

5. Guo, Y.; Chen, T. H.; Zeng, X.; Warburton, D.; Bostrom, K. I.; Ho, C. M.; Zhao, X.; Garfinkel, A. (2014). Branching patterns emerge in a mathematical model of the dynamics of lung development, *J Physiol.*, **592**, pp. 313–24.

6. Mironov, V.; Trusk, T.; Kasyanov, V.; Little, S.; Swaja, R.; Markwald, R. (2009). Biofabrication: A 21st century manufacturing paradigm, *Biofabrication*, **1**, pp. 1–16.

7. Lei, Y.; Gojgini, S.; Lam, J.; Segura, T. (2011). The spreading, migration and proliferation of mouse mesenchymal stem cells cultured inside hyaluronic acid hydrogels, *Biomaterials*, **32**, pp. 39–47.

8. Mimeault, M.; Batra, S. K. (2006). Concise review: Recent advances on the significance of stem cells in tissue regeneration and cancer therapies, *Stem Cells*, **24**, pp. 2319–45.

9. Eiraku, M.; Takata, N.; Ishibashi, H.; Kawada, M.; Sakakura, E.; Okuda, S.; Sekiguchi, K.; Adachi, T.; Sasai, Y. (2011). Self-organizing optic-cup morphogenesis in three-dimensional culture, *Nature*, **472**, pp. 51–6.

10. Rizzoti, K.; Lovell-Badge, R. (2011). Regenerative medicine: Organ recital in a dish, *Nature,* **480**, pp. 44–6.

11. Bashur, C. A.; Venkataraman, L.; Ramamurthi, A. (2012). Tissue engineering and regenerative strategies to replicate biocomplexity of vascular elastic matrix assembly, *Tissue Engineering Part B: Reviews,* **18**, pp. 203–217.

12. Trkov, S.; Eng, G.; Di Liddo, R.; Parnigotto, P. P.; Vunjak-Novakovic, G. (2010). Micropatterned three-dimensional hydrogel system to study human endothelial-mesenchymal stem cell interactions, *J. Tissue Eng. Regen. Med.,* **4**, pp. 205–15.

13. Jakab, K.; Norotte, C.; Damon, B.; Marga, F.; Neagu, A.; Besch-Williford, C. L.; Kachurin, A.; Church, K. H.; Park, H.; Mironov, V.; Markwald, R.; Vunjak-Novakovic, G.; Forgacs, G. (2008). Tissue engineering by self-assembly of cells printed into topologically defined structures, *Tissue Eng Part A,* **14**, pp. 413–21.

14. Danino, T.; Volfson, D.; Bhatia, S. N.; Tsimring, L.; Hasty, J. (2011). In-silico patterning of vascular mesenchymal cells in three dimensions, *PLoS ONE,* **6**, e20182.

15. Yochelis, A.; Tintut, Y.; Demer, L. L.; Garfinkel, A. (2008). The formation of labyrinths, spots and stripe patterns in a biochemical approach to cardiovascular calcification, *New Journal of Physics,* **10**, pp. 78–106.

16. Chinake, C. R.; Simoyi, R. H. (1997). Experimental studies of spatial patterns produced by diffusion-convection-reaction systems, *Journal of the Chemical Society-Faraday Transactions,* **93**, pp. 1345–1350.

17. Sheth, R.; Marcon, L.; Bastida, M. F.; Junco, M.; Quintana, L.; Dahn, R.; Kmita, M.; Sharpe, J.; Ros, M. A. (2012). Hox genes regulate digit patterning by controlling the wavelength of a Turing-type mechanism, *Science,* **338**, pp. 1476–1480.

18. Turing, A. M. (1952). The chemical basis of morphogenesis, *Philos. Trans. R. Soc. Lond. B Biol. Sci.,* **237**, pp. 37–72.

19. Garfinkel, A.; Tintut, Y.; Petrasek, D.; Bostrom, K.; Demer, L. L. (2004). Pattern formation by vascular mesenchymal cells, *P Natl Acad Sci USA,* **101**, pp. 9247–9250.

20. Chen, T. H.; Zhu, X.; Pan, L.; Zeng, X.; Garfinkel, A.; Tintut, Y.; Demer, L. L.; Zhao, X.; Ho, C. M. (2012). Directing tissue morphogenesis via self-assembly of vascular mesenchymal cells, *Biomaterials,* **33**, pp. 9019–9026.

21. Xu, H.; Sun, M.; Zhao, X. (2017). Turing mechanism underlying a branching model for lung morphogenesis, *Plos One,* **12**, pp. e0174946.

22. Marcon, L.; Sharpe, J. (2012). Turing patterns in development: What about the horse part?, *Curr. Opin. Genet. Dev.,* **22**, pp. 578–584.

23. Zhu, X.; Gojgini, S.; Chen, T. H.; Fei, P.; Dong, S.; Ho, C.-M.; Segura, T. (2017). Directing three-dimensional multicellular morphogenesis by self-organization of vascular mesenchymal cells in hyaluronic acid hydrogels, *Journal of Biological Engineering,* **11**, pp. 12.

24. Zhu, X.; Gojgini, S.; Chen, T. H.; Teng, F.; Fei, P.; Dong, S.; Segura, T.; Ho, C.-M. (2016). Three dimensional tubular structure self-assembled by vascular mesenchymal cells at stiffness interfaces of hydrogels, *Biomedicine & Pharmacotherapy,* **83**, pp. 1203–1211.

25. Chen, T. H. (2014). Tissue regeneration: From synthetic scaffolds to self-organizing morphogenesis, *Current Stem Cell Research & Therapy,* **9**, pp. 432–443.

26. Sant, S.; Hancock, M. J.; Donnelly, J. P.; Iyer, D.; Khademhosseini, A. (2010). Biomimetic gradient hydrogels for tissue engineering, *Can. J. Chem. Eng.,* **88**, pp. 899–911.

27. Benjamin, M.; Evans, E. J.; Copp, L. (1986). The histology of tendon attachments to bone in man, *J. Anat.,* **149**, pp. 89–100.

28. Lu, H. H.; Jiang, J. (2006). Interface tissue engineering and the formulation of multiple-tissue systems, *Adv. Biochem. Eng. Biotechnol.,* **102**, pp. 91–111.

29. Moffat, K. L.; Wang, I. N.; Rodeo, S. A.; Lu, H. H. (2009). Orthopedic interface tissue engineering for the biological fixation of soft tissue grafts, *Clin Sports Med.,* **28**, pp. 157–76.

30. Zimmerman, L. B.; DeJesusEscobar, J. M.; Harland, R. M. (1996). The Spemann organizer signal noggin binds and inactivates bone morphogenetic protein 4, *Cell,* **86**, pp. 599–606.

31. Rifas, L. (2007). The role of noggin in human mesenchymal stem cell differentiation, *J. Cell. Biochem.,* **100**, pp. 824–834.

32. Chen, T. H.; Hsu, J. J.; Zhao, X.; Guo, C.; Wong, M. N.; Huang, Y.; Li, Z.; Garfinkel, A.; Ho, C. M.; Tintut, Y.; Demer, L. L. (2012). Left-right symmetry breaking in tissue morphogenesis via cytoskeletal mechanics, *Circ Res.,* **110**, pp. 551–9.

33. Zhu, X.; Yang, Y. (2016). In *Simulation for tubular and spherical structure formation via self- organization of vascular mesenchymal cells in three dimensions*, 9th International Congress on Image and Signal Processing, BioMedical Engineering and Informatics (CISP-BMEI), Taiyuan, China, Oct. 15–17, 2016, pp. 1654–1659.

34. Zhu, X.; Yang, H. (2016). In *In-silico constructing three-dimensional hollow structure via self-organization of vascular mesenchymal cells,* IEEE 16th International Conference on Nanotechnology (IEEE-NANO), Aug. 22–25, 2016, pp. 468–471.

Chapter 6

Tuning Cellular Behaviors during Self-Organization of Cells in Hydrogel by Changing Inner Nano-Structure of Hydrogel

6.1 Introduction

The self-organizing formation of multicellular structures in vitro probably could potentially recapitulate the morphological structure and organization of native tissues [1–4]. As we know, the human body is self-assembled by numerous hierarchical tissue architectures in a natural manner, and each type of tissue is hierarchically self-assembled by modular building blocks [5], such as modular and hierarchical organizations occurred in the human body. As typical self-assembling objects, cells have the dimensions much larger than the molecules transported inside or outside the cells in the 3D matrix, and usually can be easily observed.

Self-organization of cells usually can be achieved in hydrogel system [6–8]. In vitro cell culture has always been important in the biomedical field for decades [9, 10]. With the development of various biomaterials, it is more convenient for us to construct tailored in vitro cell culture models by tuning the composition or configuration of the biomaterials for studying the physiological characteristics of cells deeply and understanding the inner mechanisms clearly [11, 12]. As described in previous chapters, the mainstream in vitro cell culture

Self-Organized 3D Tissue Patterns: Fundamentals, Design, and Experiments
Xiaolu Zhu and Zheng Wang
Copyright © 2022 Jenny Stanford Publishing Pte. Ltd.
ISBN 978-981-4877-77-0 (Hardcover), 978-1-003-18039-5 (eBook)
www.jennystanford.com

techniques mainly consists of 2D culture and 3D culture [13–15]. For the cells immersed in the culture medium in 2D culture, the surface-modified substrate usually provides an opening space for cells to attach on [16]. Choosing different properties of substrate material could induce distinguishing growth performances of cells [14, 16]. The designing of 3D matrix for inducing special performance of cells stems from the previous methodology in 2D cell culture was stated above [17–19]. A 3D cell culture using hydrogel as the 3D substrate material could construct the suitable extracellular matrix (ECM) for cells [20, 21], which could more realistically mimic native biochemical and biomechanical microenvironments compared with 2D culture [17, 19, 22]. The configuration of substrate materials for cells usually plays an important role in the research on 2D and 3D cultures, because the specifically designed functionalized biomaterials could influence the cellular behaviors via the microenvironments in multiple aspects [23, 24]. To provide a healthy and adjustable environment for cells in 3D culture requires hydrogels with high hydrophilicity, porosity, variable mechanical properties, and good biocompatibility [25–27]. To meet these requirements for the 3D hydrogel scaffolds, numerous natural and synthetic hydrogels including materials such as dextran, hyaluronic acid, alginate, poly (ethylene glycol), and poly (vinyl alcohol) [28–31] have been fabricated and investigated. Usually the physical and chemical properties of hydrogels can be regulated to satisfy the requirements of cell growth, such as cellular adhesion, migration, proliferation, and differentiation [29, 32, 33]. Some hydrogel materials showed potential in regulating cell growth parametrically and quantitatively [34, 35].

Further, improved methods of fabricating such hydrogel matrices can help direct single-cell and collective cellular behaviors, by which researchers can pierce into the cell–ECM interactions more conveniently [28, 36–38]. For example, RGD peptides [39] were known as the anchors of cellular adhesion, and they function as the connection between scaffold polymers and integrin spanning the cell membrane. Integrin clustering plays an important role in the activation of the signal transduction pathway that mediates cellular activities such as reorganization of intracellular cytoskeleton, regulation of the growth factors, and control of ion channels [40].

The spatial distribution of RGD peptides in the scaffolds can influence cell adhesion dynamics and motility. Researches in 2D matrices have shown RGD spatial distribution may affect cell adhesion on the nanoscale [41, 42]. However, the impact laws of the specific RGD distribution on the collective cell behaviors in 3D hydrogels have not been intensively and fully characterized and analyzed, which hinder the in-depth exploration for the underlying mechanism during the cellular self-organization.

Dextran hydrogel [43–47], as one of the semisynthetic hydrogels, can be modified to satisfy the requirements of directing cell growth. Dextran is a linear polysaccharide consists of α-1,6 linked D-glucopyranose, and it can be modified with diverse functional groups for promoting further cell behaviors [48, 49]. It can help the fabrication of variants of dextran hydrogel as 3D culture models with chosen mechanical and biochemical properties and the microenvironment better mimicking in vivo ECM [49–52].

In this study, we fabricated the maleimide-dextran (MD) hydrogel with homogenous and clustering distributions of RGD peptides, and employed the 3D dextran hydrogels parametrically designed with homogenous and clustered RGD compositions to explore the flexible and quantitative regulation of the substrate materials influencing the collective cellular behaviors. The growth characteristics of NIH-3T3 fibroblasts (NIH stands for National Institutes of Health) and C2C12 cells within 3D dextran hydrogel with homogenous composition were experimentally studied here. The effect of the hydrogel materials on cellular behaviors has been discussed relative to that on 2D substrates. Further, we fabricated the dextran hydrogels with four different clustering degrees of RGD compositions and measured their cell-adhesion efficacy. The property of the RGD-homogenous hydrogel was characterized through scanning electron microscope (SEM) imaging and rheological analysis. The cytotoxicity assay, survival rate analysis, and proliferation measurement of cells were quantitatively conducted for demonstrating the biocompatibility of MD hydrogel. The RGD-clustered hydrogel was quantitatively assessed via measuring the hydrogel's cell-adhesion efficacy, observing the evolutionary multicellular morphology and the distribution of F-actins inside cells in 3D. The results showed that RGD-clustered hydrogel has the advantage of enhancing cell elongation and the connection or aggregating degree of cells, compared to the hydrogel with homogenous RGD. Varying RGD-clustering degree in hydrogels could serve as a stable approach for modulating the cellular growth behaviors. This study on the clustered RGD composition in 3D dextran hydrogel provides useful information for quantitatively designing the tailored hydrogel system and opens a new avenue for quantitatively regulating the cell–biomaterial interactions during the cellular self-organization and tissue morphogenesis.

6.2 Materials and Methods

6.2.1 3D Dextran Hydrogel

The dextran hydrogel (Cat No: FG91-1, Cellendes, Reutlingen, Germany) was prepared with deionized (DI) water, 10-fold concentrated buffers (10 × CB, pH 5.5), MD (thiol-reactive polymer), and polyethylene glycol peptide conjugate (CD-Link, thiol-containing crosslinker). It is a quick-forming hydrogel that can be fabricated in approximately 3–5 min. The time of gel formation is mainly dependent on the concentration of thioether and the pH value of the buffer system [53]. The mechanical properties of hydrogel, such as stiffness and viscosity, correspond to the proportion of gel components. In this study, the crosslinking process happened between thiol-reactive MD and thiol-containing CD-Link (Figure 6.1a). Gel stiffness was positively correlated to the concentration of reacted maleimide groups, and this concentration was defined as the crosslinking strength in this study. Additionally, the thiol-containing RGD peptides (Cat No: 09-P-001, Cellendes, Reutlingen, Germany) were used to functionalize the MD for cell adhesion, and thioglycerol (Cat No: T10-3, Cellendes, Reutlingen, Germany) was used to maintain the equal final concentration between maleimide and thiol groups (Figure 6.1a). Such 3D dextran hydrogel can be easily degraded by the dextranase (Figure 6.1b). The sequence of RGD peptide is Acetyl-Cys-Doa*-Doa-Gly-Arg-Gly-Asp-Ser-Pro-NH$_2$ (*: Doa: 8-amino-3,6-dioxaoctanoic acid).

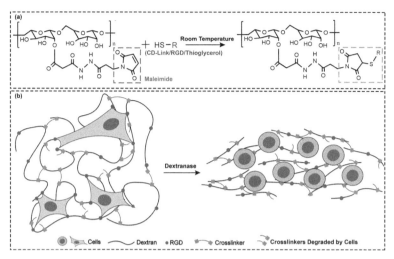

Figure 6.1 Schematic diagrams of fabrication and degradation of 3D homogenous mal-dextran hydrogels. (**a**) Maleimide–thiol reaction formula: maleimide groups were on dextran and thiol groups (-SH) were on CD-Link (thiol-containing crosslinker), RGD peptides, and thioglycerol. (**b**) Degradation of the formed dextran hydrogel.

6.2.2 Cell Preparation

NIH-3T3 fibroblasts and C2C12 cells were cultured on the substrate surface of the 100 mm × 20 mm cell culture dish (Nest Biotechnology Co. Ltd., Wuxi, China). The components of the culture medium were 1% penicillin–streptomycin solution (HyClone, GE Healthcare Life Sciences, South Logan, UT, the USA), 10% fetal bovine serum (EVERY GREEN, Tianhang Biotechnology Co. Ltd., Zhejiang, China), and 89% DME/F-12 (HyClone, GE Healthcare Life Sciences, South Logan, UT, the USA). The inside environment of the cultivator was 37 °C and 5% CO_2. Cells were plated at 1×10^6 in each dish, and were passaged every three days. In this study, the passage number of NIH-3T3 fibroblasts was P7–P10, and that of C2C12 was P14–P17. To harvest cultured cells from the substrate, replace the culture medium with 2 mL of phosphate-buffered saline (PBS) (HyClone, GE Healthcare Europe GmbH, Freiburg, Germany) and 1 mL of trypsin 0.25% (1X) solution (HyClone, GE Healthcare Life Sciences, South Logan, UT, the USA). Cells were incubated in the cultivator for 4 min before being detached from the substrate. Then, the detached cells were transferred into a 15-mL centrifuge tube, centrifuged (3T3, 1000 r/min, 3 min and C2C12, 1000 r/min, 6 min), resuspended in fresh culture medium, and counted with a Metallized Hemacytometer (HAUSSER SCIENTIFIC, Horsham, PA, the USA).

6.2.3 RGD-Homogenous Hydrogel Fabrication

To obtain the precursor solution, DI water, 10 × CB (pH 5.5), MD, RGD peptides, and thioglycerol were added into a reaction tube in proportion and mixed thoroughly. Then, the precursor solution was incubated for 5–10 min in room temperature for a complete reaction (Figure 6.1a). CD-Link was placed onto the bottom of the wells of a 96-well plate. The cell suspension was added into the precursor solution and mixed evenly. The final cell density in the gel was set at 5 k/µL. Then, the cell-containing precursor solution was transferred into the wells containing CD-Link and mixed two times quickly and pliably. During this operation, it was critical to avoid air bubbles, which would influence the later observation and imaging. The hydrogel was completely formed in 3–5 min at room temperature. The sample was covered with fresh culture medium and incubated in the cultivator. The medium was renewed after cultivation of 2 h. Medium was changed every two days during cultivation. The volume of each hydrogel sample for cell culture was always maintained at 30 µL, and the reagents were kept on ice. As the control, 3T3 and C2C12 cells were also culture on tissue culture polystyrenes (TCPS) respectively.

6.2.4 RGD-Clustered Hydrogel Fabrication

The fabrication method of homogenous RGD-functionalized dextran hydrogel has been given in Section 6.2.3. This section gave the improved method for regulating the clustering rate (or homogenous level) of RGD distribution in dextran hydrogel. When preparing the precursor solution, MD was divided into two parts for the mixture (Figure 6.2). DI water, 10 × CB (pH 5.5), the first part of MD, and RGD peptides were added into a reaction tube in proportion, mixed thoroughly, and incubated for 5–10 min in room temperature.

After the reaction was completed, the second part of MD and thioglycerol were added into the reaction tube, mixed thoroughly, and incubated for 5–10 min in room temperature. Four different cases were tested, and the related parameter values were listed in Table 6.1. The final cell density in the gel was set at 2.5 k/µL. The next steps for hydrogel fabrication followed the contents in Section 6.2.3.

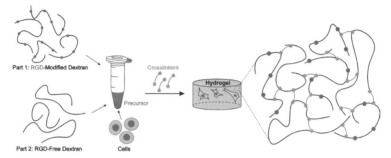

Figure 6.2 Illustration of the method for fabricating 3D dextran hydrogels with different clustered RGD compositions.

Table 6.1 Four different RGD distributions were performed for 3D dextran hydrogel. The levels of RGD clustering decreased from Group 1 to Group 4

Parameter	Group 1	Group 2	Group 3	Group 4
RGD concentration per gel (µM)	50	50	50	50
Total amount of RGD per gel (nmol)	1.5	1.5	1.5	1.5
% Mal-dextran reacted with RGD	16.7	33.3	66.6	100
RGD clustering ([mmol RGD]/[mmol maleimide group in the first part])	0.1	0.05	0.025	0.017

6.2.5 SEM Imaging

The hydrogel samples were imaged with a SEM (Sigma-500, Zeiss, Oberkochen, Germany). The samples were cut out to indicate their internal surfaces. The working voltage was set at 10.0 kV, and the working distance was 9.00 mm. Gel images magnified 5000 and 20,000 times have been obtained.

6.2.6 Rheology Measurement

The elastic modulus (G') and viscous modulus (G") of hydrogel with three different crosslinking strengths were measured with a plate-to-plate rheometer (Kinexus Pro, Malvern, the UK). The volume of each hydrogel sample was 90 µL, and the gap distance was set at 0.2 mm. The complex shear strain was 1%, and the frequency ranged from 0.1 Hz to 10 Hz with the temperature inside the humid hood set at 37 °C.

6.2.7 Live/Dead Test

A Live/Dead Viability/Cytotoxicity Kit (Cat No: L3224, Thermo Fisher Scientific, Eugene, OR, the USA) was used for testing cellular viability. To get the live/dead dye solution, 3.6 µM ethidium homodimer-1 and 2.8 µM calcein acetoxymethyl ester (calcein-AM) solutions were obtained by dissolving them in 1 mL of DME/F-12. The medium was moved out from the well, and the sample was washed with PBS two times. Solution (150 µL) was added into the well, and the sample was incubated in dark at 37 °C for 25 min. Then, the sample was washed with PBS three times, 5–10 min each time, and observed and imaged under the laser with PBS left in the well. Cells were observed by an inverted microscope (Olympus IX73, Olympus Corporation, Tokyo, Japan), and the images were taken with a digital sCMOS camera (HAMAMATSU C11440-42U30, Hamamatsu Photonics K.K., Hamamatsu City, Japan). With the help of ImageJ (Oracle Corporation, Redwood Shores, CA, USA), we counted the number of living and dead cells in the images. In 3D culture, the sample numbers of survival rates of 3T3 on day 0, day 3, day 6, and day 9 were 9, 7, 7, and 11; the sample numbers of survival rates of C2C12 on day 0, day 3, day 6, and day 9 were 4, 7, 4, and 10, respectively. In 2D culture, the sample numbers of survival rates of 3T3 on day 0, day 3, day 6, and day 9 were 5, 5, 5, and 5; the sample numbers of survival rates of C2C12 on day 0, day 3, day 6, and day 9 were 5, 5, 5, and 5, respectively. We conducted the calculation for the mean and standard deviation of the data.

6.2.8 Bright Field Imaging

With an inverted microscope and the digital sCMOS camera, we took the images of 3T3 and C2C12 under the bright field. Specially, the filopodia and lamellipodia of cells in 2D and 3D were imaged, and we counted the amount of filopodia of 3T3 in 2D, 3T3 in 3D, C2C12 in 2D, and C2C12 in 3D, respectively, and the cell sample numbers were 57, 35, 32, and 53, correspondingly. Additionally, in this study, cellular behaviors such as spreading, sprouting, and both were recognized as the symbol of cell adhering.

6.2.9 F-actin Staining

Alexa Fluor 546 phalloidin (Cat No: A22283, Invitrogen, Thermo Fisher Scientific, Eugene, OR, the USA) was used for F-actin staining. The medium was moved out from the well, and the sample was washed with PBS three times. Cells were fixed with 4% formaldehyde (Cat No: 28908, Thermo Scientific, Rockford, IL, the USA) solution in PBS in the dark at room temperature for 45 min, and the sample was washed with PBS three times. Then, cells were extracted with a solution of 0.1% Triton X-100 (Cat No: X100-5ML, SIGMA-ALDRICH, St. Louis, MO, the USA) in PBS for 5 min. Then, the sample was washed two or more times with PBS. One kit of Alexa Fluor 546 phalloidin was dissolved in 1.5 mL of methanol (Cat No: M116128-1L, Aladdin Industrial Corporation, Shanghai, China) to get the 40× stock solution. Then, 10 µL of stock solution was diluted in 200 µL of PBS for each well to be stained. To reduce the nonspecific background, 1% bovine serum albumin (BSA) (Cat No: 37525, Thermo Scientific, Rockford, IL, the USA) was added into the staining solution. The staining process lasted 20 min in dark at room temperature. Then, the sample was washed with PBS at least three times.

6.2.10 DAPI Staining

The fixed and permeabilized sample was washed with PBS first. To get the dye solution, 4′, 6-diamidino-2′-phenylindole, dihydrochloride (DAPI) (Cat No: 62247, Thermo Scientific, Dreieich, Germany) stock solution was diluted to 138 ng/mL in PBS. Then, 300 µL of DAPI staining solution was added into each well, and the sample was incubated in dark at room temperature for 5 min. Then, the sample was washed with PBS at least three times.

6.2.11 LSCM Imaging

The F-actin and DAPI stained samples were imaged by a laser scanning confocal microscope (LSCM) (LSM-710, Zeiss, Oberkochen, Germany). Samples were transferred onto the microscope cover glass (Cat No: 12-541-B, Fisher Scientific, Waltham, MA, the USA) and soaked with PBS. Z-axis accuracy was set at 0.56 μm with an image model of 16 bit and pixel dwell time of 0.79 μs.

6.2.12 Nucleus Circularity Measuring Method

The ellipses with the circularity ranging from 0.1 to 1 at an interval of 0.02 were drawn by MATLAB (Matlab Online R2019b, MathWorks Inc., Natick, MA, the USA, trial version). These ellipses were then compared with the nucleus in the image of DAPI staining results. Once finding the ellipse with the closest shape to the measured nucleus, the circularity value of that ellipse was taken as the measured nucleus's circularity value. By using the same method, the circularity of the nucleus in images was estimated, and the mean and standard deviation of data was then calculated and charted. The cell sample numbers used for assessing nucleus circularity of 3T3 in 2D, 3T3 in 3D, C2C12 in 2D, and C2C12 in 3D were 45, 45, 45, and 45, respectively. Circularity was defined as follows:

$$\text{Circularity} = \frac{4\pi S}{L^2} \tag{6.1}$$

where S is area and L is perimeter of single nucleus.

6.2.13 Gel Degradation

The medium was moved out from the microwell. The sample was covered with 300 μL of a 1:20 dilution of dextranase (Cat No: D10-1, Cellendes, Reutlingen, Germany) in culture medium and incubated at 37 °C for 30 min. Gels could be dissolved faster if they were cut into pieces. After the degradation of the gel, the cell suspension was centrifuged, and cells were resuspended in fresh culture medium. We counted the number of cells with a metallized hemacytometer. The gel sample numbers for counting 3T3 cells on day 3, day 6, and day 9 were 3, 3, and 3; the gel sample numbers for counting C2C12 cells on day 3, day 6, and day 9 were 3, 3, and 3.

6.2.14 Data Statistics

The data was presented by mean ± standard deviation (Mean ± SD). Two-sample Student's T-Test was used to analyze the significant difference of the data in Origin software (OriginPro 2018 v9.5 64-bit, OriginLab Corporation, Northampton, MA, the USA, trial version). The upper limit value of significance level was set as $p < 0.05$. All the experiments were repeated at least three times.

6.3 Results

6.3.1 Microgeometry and Rheological Properties of Dextran Hydrogel

We imaged the microgeometry of the 3D dextran hydrogel by SEM. The sample was cut out to image its internal surfaces. The results showed that the internal surface of the gel was pleated (Figure 6.3a). It indicated that 3D dextran hydrogel can provide a rough contact surface for cells in it. Some multipore structures were presented in Figure 6.3b. The elastic modulus (G', or called storage modulus), viscous modulus (G'', or called loss modulus) of the dextran hydrogel with different crosslinking strengths were measured with a plate-to-plate rheometer at 37 °C. Results showed that the G' and G'' separately settled on the different orders of magnitude over the entire ranging of measured frequencies (0.1–10 Hz) (Figure 6.3c), and the value of G' and G'' is correlated to the mechanical properties of hydrogel. The average value of G'/G'' was lower than 0.1, which indicated that the elastic property of dextran hydrogel was more pronounced than its viscosity. Dextran hydrogel, used in this study, can be fabricated with different stiffness by allocating the proportion of MD, CD-Link, and RGD peptides. The crosslinking strength of dextran hydrogel was defined as the concentration of maleimide groups from dextran crosslinked by thiol groups from CD-Link.

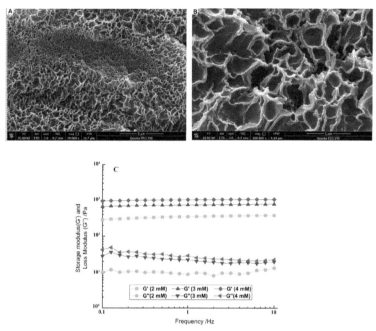

Figure 6.3 SEM images and viscoelasticity of the 3D homogenous dextran hydrogels. The main parameters of dextran hydrogel were crosslinking strength = 2 mM/ 3 mM/ 4 mM, and RGD = 300 μM. (**a**) Image of 3D dextran hydrogel with 20,000 times magnification under SEM. (**b**) Image of 3D dextran hydrogel with and 100,000 times magnification under SEM. (**c**) The storage modulus (G′) and loss modulus (G″) of dextran hydrogel measured by a plate-to-plate rotational rheometer.

6.3.2 Cellular Morphology and Behaviors in RGD-Homogenous Dextran Hydrogel

Due to the different dimensions of the substrate provided by substrate materials, cells showed a distinguishing configuration of patterns in 3D dextran hydrogel and a 2D petri dish. 3T3 fibroblasts and C2C12 cells usually self-assembled into the plump and stereo 3D shapes, while the shape of cells in 2D tended to be more flattened over two weeks of cultivation, as shown in Figure 6.4. The cellular sprouting and spreading are the representative cellular behaviors reflecting cell–matrix interactions. With the extension of culturing time, cells occupied more space in or on the substrate material by continuous proliferation, connectivity, and migration. In 3D dextran hydrogel, patterns of 3T3 and C2C12 were exhibited as distributed structures, i.e. isolated multicellular clumps were cross-connected by a path of chains composed of cells (Figure 6.4, 3T3-3D and C2C12-3D). In the 2D petri dish, patterns of cells usually kept an even cellular distribution and more random directions of cellular polarization. During the entire cultivation, the locations of 3T3 and C2C12 were almost isotropic (Figure 6.4, 3T3-2D and C2C12-2D).

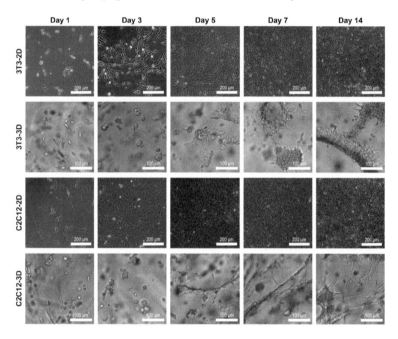

Figure 6.4 Bright-field images of 3T3 and C2C12 cells in a 2D petri dish and 3D RGD-homogenous dextran hydrogel from day 1 to day 14. The main parameters of dextran hydrogel were crosslinking strength = 2 mM and RGD = 300 μM.

Despite the same culturing conditions, 3T3 and C2C12 cells in 3D dextran hydrogel also exhibited distinguishing growth phenotypes. For these cells, the polarized state could be observed approximately 12–24 h after implantation initially. However, C2C12 cells presented greater degree of autologous stretching than 3T3 (Figure 6.4, day 5 to day 14, 3D). In addition, their rates of sprouting were distinct. For 3T3, it usually took 2–3 days to sprout mostly. Under the same conditions, that occurred on C2C12 after about 1–2 days of culturing.

In our observation, filopodia and lamellipodia were the two main cellular growing structures during sprouting and spreading. Lamellipodia are cytoskeletal protein actin projections on the front edge of the cell. Filopodia are slender cytoplasmic projections extending from the front edge of lamellipodia in migration cells. In 3D, most 3T3 and C2C12 cells were usually observed growing out some slender filopodia, while their lamellipodia were relatively inconspicuous in the hydrogel (Figure 6.5a). In 2D, both 3T3 and C2C12 cells had attachment, with their filopodia and lamellipodia adherent on the surface of petri dishes (Figures 6.5a). The polygon-shape of cells in 2D was mainly associated to the amount and morphology of their cellular lamellipodia.

By comparison, the amount of filopodia of cells was changed when transferred from petri dishes into dextran hydrogel. Once spread in 3D hydrogel, the amount of filopodia of 3T3 was larger than that in the 2D petri dish (Figure 6.5b). In contrast, after being transferred from 2D to 3D, the amount of observable filopodia of C2C12 got reduced (Figure 6.5b). The main morphological characteristics of self-organization were cell clumps and cellular connections with a bigger scale. 3T3 cells tended to proliferate to form clumps. With continuous proliferation, the volume and number of 3T3 cell groups gradually increased. Connections between the clumps of 3T3 cells usually rose after that. The process started with a single cell to a mass; then, the cells were attracted to and connected with each other by growing out cells similar to a bridge (Figure 6.4, 3T3-3D). Both multicellular clumps and connecting structures were composed of dense cells. In contrast, C2C12 preferred to stretch itself longer and connect with each other first, and long chains or mesh structures were usually observed from C2C12 in 3D dextran hydrogel (Figure 6.4, C2C12-3D). In the process of connection, proliferation also occurred. After culturing for more than one week, it was common to observe that C2C12 cells of clumps extended around and connected with each other by slender bundles of stretched cells.

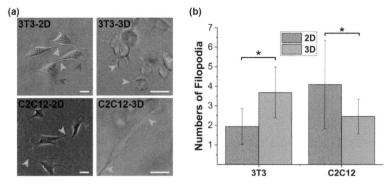

Figure 6.5 Lamellipodia and filopodia of cells in 2D dish and 3D dextran hydrogels. (**a**) Lamellipodia (indicated by pink arrows) and filopodia (indicated by green arrows) of 3T3 and C2C12 in 2D and 3D. (**b**) Numbers of filopodia of cells in 2D petri dishes and 3D dextran hydrogel. The data were presented by mean ± SD; *$p < 0.05$ versus the corresponding 2D control samples. All the scale bars are 20 μm.

F-actin and DAPI staining were conducted on cells in the 2D petri dish and 3D dextran hydrogel on day 14, respectively. Images of cellular nucleus and microfilament taken under the LSCM exhibited the multicellular structures of 3T3 and C2C12 in 2D and 3D, as shown in Figure 6.6a. As different substrate materials, dextran hydrogel, and the petri dish surface contributed distinguishing cytoskeleton and self-organization characteristics to the cells. Both 3T3 and C2C12 in 3D dextran hydrogel tended to grow into multicellular clumps with a densely packed nucleus, and the shape of their nucleus was influenced in the hydrogel materials. That is, the nucleus shape of 3T3 and C2C12 in hydrogel presented lower circularity than that of cells in dishes (Figures 6.6b and c). In addition, cells usually spread out on the bottom surface of the petri dish with the microfilaments incompact from each other. In contrast, 3T3 and C2C12 in dextran hydrogel presented more agglomeration and connectivity with the microfilaments squeezed tightly with each other (Figure 6.6a). Compared to the results in 2D, the growth of 3T3 and C2C12 in 3D dextran hydrogel was featured with more 3D interleaving structures. To be more specific, during 14 days of cultivation, 3T3 grew out many banded structures with numerous cells gathered and self-organized together and distributing in the hydrogels. Some doughnut-shaped holes with various sizes were usually observed in the multicellular connection structures of 3T3 (Figure 6.6a, 3T3-3D). For C2C12 cells in 3D dextran hydrogel, the staggered 'truss' structures that can be found are typical of them. C2C12 cells in the same path tended to connect into a slender long chain, followed with proliferation. C2C12 cells in such chain structures tended to exhibit lower circularity of the cell nucleus (Figure 6.6a, C2C12-3D).

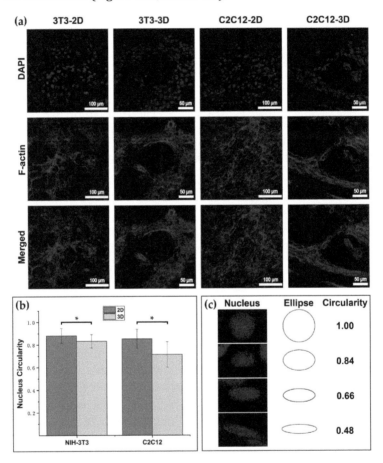

Figure 6.6 Cytoskeletal staining and analysis for cells in 2D and 3D on day 14. **(a)** DAPI and F-actin staining of 3T3 and C2C12 cells in 2D and 3D culture. **(b)** Circularity of cellular nucleus in 2D petri dishes and 3D dextran hydrogel. The data was presented by mean ± SD; *p < 0.05 versus the corresponding 2D control samples. **(c)** Demonstrations of estimating the circularity of cell nucleus.

6.3.3 Cell-Adhesive Efficacy of RGD Clustering Dextran Hydrogels

3D dextran hydrogels with four different RGD clustering compositions have been fabricated and tested for culturing NIH-3T3 fibroblasts and C2C12 cells for one week, respectively. The proportions of the mal-dextran that reacted with RGD (abbreviated to proportions of dextran with RGD) for four cases are 16.7%, 33.3%, 66.7%, and 100%, respectively. According to the images under the bright field, C2C12 and 3T3 cells could normally spread and sprout in both RGD-homogenous and RGD-clustered dextran hydrogels (Figure 6.7a C2C12-day 1 and Figure 6.7b 3T3-day 1). With the extension of culturing time, C2C12 cells gradually became more elongated and connected into the long-branch structures (Figure 6.7a C2C12-day 4 and C2C12-day 7). Then, 3T3 cells tended to aggregate into clumps both in RGD-homogenous and RGD-clustered dextran hydrogels (Figure 6.7b 3T3-day 4 and 3T3-day 7).

We counted the adhered 3T3 and C2C12, respectively, on day 1 to access the initial cell adhesive efficacy of 3D dextran hydrogels with different clustered RGD compositions. The results showed that for both 3T3 and C2C12, cells showed relatively higher average adhesion rates in 3D hydrogels with more homogenous RGD composition (Figures 6.8a and b). Additionally, the adhesion rates of C2C12 cells were higher than those of 3T3 cells generally in most cases. It can be observed under the bright field, cells could normally spread, sprout, and aggregate in these four cases. The specific performances varied on different cells, which have been partly mentioned in the homogenous case above. Further, the phalloidin staining results showed that cells with higher-level extension or elongation tended to appear as 'well-formed' F-actin skeletons, which means curves of intracellular microfilaments fit well with the postures of the cellular body (Figure 6.8c). The maximum length (ML) of spreading C2C12 cells was measured on day 4 in groups 1–4 for further estimating the performance of the hydrogel inducing and maintaining the cellular spreading. The results showed that the ML of C2C12 cells in the RGD-clustered cases was significantly higher than that in the homogenous case (Figure 6.8d), and that differences became more significant when the RGD clustering rate increased.

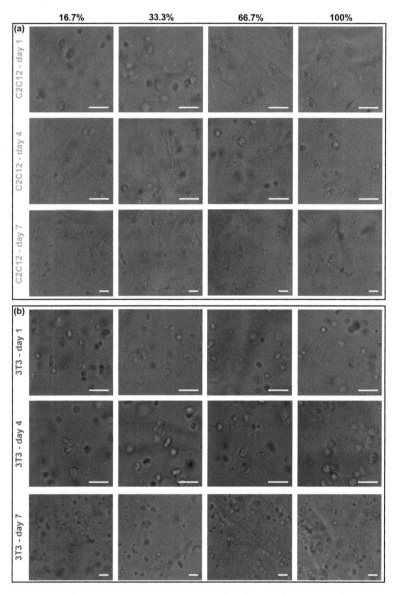

Figure 6.7 The images of C2C12 and 3T3 cells in four different RGD clustering cases under the bright field. The main parameters of dextran hydrogel were crosslinking strength = 2 mM and RGD = 50 μM. (a) C2C12 cells, and (b) 3T3 cells. All the scale bars, 50 μm. Cells were observed by an inverted microscope, and the images were taken with a digital sCMOS camera. The proportions of the mal-dextran reacted with RGD for four cases are 16.7%, 33.3%, 66.7%, and 100%, respectively.

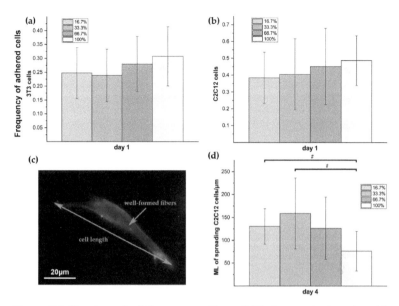

Figure 6.8 The impact of the composition of 3D dextran hydrogels with clustered and homogenous RGD distribution on initial cell adhesion. The main parameters of dextran hydrogel were a crosslinking strength = 2 mM and RGD = 50 μM. (**a**) Adhesion rates of 3T3 cells on day 1 in groups 1–4. (**b**) Adhesion rates of C2C12 cells on day 1 in groups 1–4. (**c**) The well-formed fibers of adhered C2C12 cells were stained by phalloidin. (**d**) The ML of spreading C2C12 cells was measured on day 4 in groups 1–4. The measurement was conducted by ImageJ software. The data were presented by mean ± SD; #$p < 0.01$ versus the corresponding 3D homogenous samples.

6.3.4 Cell Spreading, Elongation, and Connection in RGD-Clustering Dextran Hydrogels

The elongation and polarization of C2C12 cells and protrusion and aggregation of 3T3 cells were observed in Figures 6.9a and b. With the extension of culturing time, C2C12 cells came into connections, and the connected multi-cells also showed polarized structures. The amount of C2C12 cells connecting tightly was counted from lengthways (parallel to the polarization) and breadthwise directions (vertical to the polarization) respectively, which were denoted as L-connected and B-connected cells, correspondingly (Figure 6.10b). The results showed that the amount of L-connected C2C12 increased with the increase of the homogenous level of RGD distribution, and the values were different significantly among the four groups (Figure 6.10e).

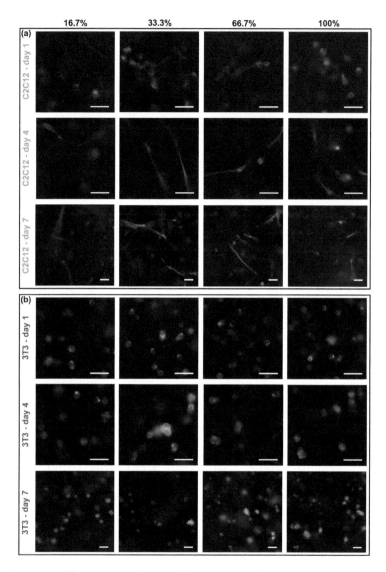

Figure 6.9 The impact of four different RGD clustering cases on the morphology of F-actin in cells over time. The main parameters of dextran hydrogel were crosslinking strength = 2 mM and RGD = 50 μM. (**a**) Morphology of F-actin in C2C12 cells. (**b**) Morphology of F-actin in 3T3 cells; cells were observed by an inverted microscope, and the images were taken with a digital sCMOS camera. The proportions of the mal-dextran reacted with RGD for four cases are 16.7%, 33.3%, 66.7%, and 100%, respectively. All the scale bars, 50 μm

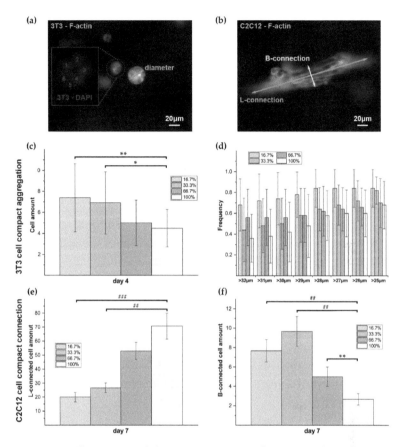

Figure 6.10 The impact of the composition configuration of 3D dextran hydrogels with clustered and homogenous RGD distribution on cellular collective behaviors. The main parameters of dextran hydrogel were crosslinking strength = 2 mM and RGD = 50 μM. (**a**) The measuring scheme for cell amount of aggregated 3T3 cells and diameter of big 3T3 cell clumps. (**b**) The measuring scheme for cell amounts of L-connected and B-connected C2C12 cells. (**c**) Cell amount of aggregated 3T3 cells on day 4 in groups 1–4. (**d**) Frequency of 3T3 cells growing into big clumps on day 7 in groups 1–4. (**e**) Cell amount of L-connected C2C12 cells on day 7 in groups 1–4. (**f**) Cell amount of B-connected C2C12 cells on day 7 in groups 1–4. The measurements were conducted by ImageJ software. The data were presented by mean ± SD; *$p < 0.05$, **$p < 0.03$, ##$p < 0.003$, and ###$p < 0.001$ versus the corresponding 3D homogenous samples.

However, for the B-connected C2C12 cells, the average values were reversed in general, except that there were relatively more B-connected C2C12 cells in the 33.3% RGD clustered dextran hydrogel than that in the 16.7% case (Figure 6.10f). 3T3 cells tended to aggregate into spherical clumps, which were observed typically in all the four different RGD-clustered 3D dextran hydrogels. The amount of compactly aggregated 3T3 cells was counted on day 4 (Figure 6.10a), and the results showed that the average values increased with the growing of the clustering level of RGD distribution. Especially, the amount of compactly aggregated 3T3 cells in the 16.7% and 33.3% RGD clustered dextran hydrogels was significantly higher than those in the homogenous case (Figure 6.10c). Such multicellular clumps grew into bigger scales continuously during cultivation. The frequencies of 3T3 cells growing into big clumps were shown in Figure 6.10a (day 7), and the results showed that 3T3 cells grew into bigger clumps in RGD clustering dextran hydrogels (Figure 6.10d).

6.4 Discussion

6.4.1 Fundamental Comparison on Cellular Morphology and Behaviors in 2D Petri Dishes and 3D Dextran Hydrogel

The MD hydrogel was used to investigate the evolution of multiplication and self-organization of NIH-3T3 fibroblasts and C2C12 cells in a 3D ECM compared with the same type of cells in 2D petri dishes. Multipore structures are ubiquitous in gels and work for sufficient flow of the culture medium (Figure 6.3b). It indicated that such hydrogel can help provide a hydrophilic ECM for soft tissues, which usually require a high water-containing environment. When 3D dextran hydrogel (RGD peptides are homogenous) and a 2D petri dish work as substrate materials, the differences between them could be highlighted by the distinguishing growth of phenotypes of cells. The results of NIH–3T3 fibroblasts and C2C12 cells in 2D petri dishes and 3D dextran hydrogel were compared, and it was found that the 2D substrate and 3D dextran hydrogel stimulated the cells in different manners and the cells responded with different behaviors, which led to the growth of different phenotypes in the corresponding matrices. Additionally, 3T3 and C2C12 cells may interact with the dextran hydrogel in different modes. Hence, they behaved differently and formed distinguished multicellular structures.

3D dextran hydrogel and a 2D petri dish conduct the mechanical stresses on cells distinguishingly, which could influence the cellular spreading and nucleus shape differently. Once settling onto the 2D dish substrate, cells formed attachment gradually, and their microfilaments were almost unconstrainedly and randomly distributed, which allowed the cells to spread freely. When transferred into hydrogel, cells became constrained and their cytoskeletons were under the stress from the surrounding microenvironment. Such stimulus influenced cells' decisions to spread or to adjust to the suitable cellular shape such as spheres. Once spread and sprouted out, the amount of filopodia of 3T3 in 3D dextran hydrogel was larger than that in the 2D petri dish, which indicated that the hydrogel enhanced the emergence of the filopodia of 3T3 cells; meanwhile, the results of C2C12 were the opposite, which indicated that the hydrogel hindered the emergence of the filopodia of C2C12 cells (Figure 6.5). Therefore, such dextran-based biomaterial can potentially provide a platform for distinguishing between 3T3 and C2C12 or even other cells. The dynamic cytoskeletal activities not only led to the variation in cellular morphology, but also caused the change of nucleus shape (Figures 6.6b and c). The decrease in the circularity of cellular nuclei, as shown in Figure 6.6b, was possibly due to the conduction of stress from the surrounding cellular cytoskeletons. The cellular cytoskeletons suffered the stress because of cell–matrix interactions such as the cell elongating itself in the elastic dextran hydrogel.

3D dextran hydrogel may help generate the locally inhomogeneous distribution of cytokines, which could induce diversified self-organized multicellular structures in such substrate material. Once transferred whether into petri dishes or dextran hydrogels, 3T3 and C2C12 cells were uniformly distributed in the 2D or 3D space (Figure 6.4, day 1). The cells in 3D dextran hydrogel gradually grew into the structures with an uneven cell distribution (Figure 6.4, 3D and Figure 6.6a, 3D). That process was accompanied with cellular proliferation, migration, and apoptosis. In contrast, the distribution of cells in 2D dishes was usually did not change obviously during the entire cultivation (Figure 6.4, 2D). Although both petri dishes and hydrogel can provide enough adhesion for cellular migration, the cells in them received different biochemical induction. For cells in 2D, at least half of the cytomembrane was exposed to the culture medium directly, and cytokines secreted by cells would quickly dilute into the surrounding medium, compared with conditions in hydrogel that has restriction for the cytokine diffusion. Therefore, it was difficult for cells in 2D dishes to decide where to move or which cell to connect with by distinguishing the cytokines coming from almost all 2D directions. It may cause the appearance that the patterns of cells in 2D dishes seemed to be uniform, because cells move in random directions. Therefore, the distribution of cells on 2D surface can be kept nearly uniform and constant. For cells in 3D dextran hydrogel, because of the retarding effect of the hydrogels on molecular diffusion [52, 54], the cytokines secreted by cells would not completely and immediately dilute into entire extra-environment, but spread into the surrounding 3D space and would get weaker gradually. Therefore, cells in hydrogel received cytokines with different strengths from different directions. Attracted by these recognizable cytokines, cells can decide where to spread and which cells to connect with. It may cause that the cells in 3D hydrogel grew into structures with an uneven cell distribution. The growth performance of the cells reflected the characteristics of interaction between cells and 3D dextran hydrogel compared with that on 2D substrates.

6.4.2 Cellular Morphology and Behaviors in RGD-Homogenous and RGD-Clustered Dextran Hydrogel

Inside the 3D hydrogel, RGD peptides function as the connection between scaffold polymer and integrin on cells. The spatial distribution of RGD peptides in the scaffolds can usually influence cell adhesion dynamics and motility. The clustering level of RGD peptides on dextran polymer alters the available amount of adhesive sites and the cellular adhesion strength on the specific position, which could finally determine the adhesion efficacy of cells on the polymers. Our data indicated that the 3D hydrogels with different RGD clustering levels gave different effects on the behaviors of both 3T3 and C2C12 cells in each growth stage (day 1, day 4, and day 7). In the case of 50 μM RGD, the more homogenously the RGD distributes, the more chances for cells to access the attaching sites. Further, in the initial stage just after implantation, the cells may prefer to probe the adjacent RGD anchor for temporary attachment. That may be the partial reason for the 3T3 and C2C12 showing higher initial adhesion rates in homogenous hydrogels on day 1 (Figures 6.8a and b). After day 1, the increasing hunger of cells for spreading and migration, especially for C2C12, cannot be satisfied by the sporadic adjacent RGD anchors in the homogenous case anymore. Meanwhile, too many initial cell–matrix conjunctions may reversely become the barriers for the further cellular contraction and elongation during migration in the next stage. The reasons for C2C12 cells that elongated longer in the highly RGD-clustered hydrogel than that in the homogenous case on day 4 could be explained below. The cell–matrix adhesion strength and the actin stress fiber formation usually depend on the characteristics of RGD ligand presentation, and previous efforts showed that cells on the 2D substrate with particularly clustered RGD tended to exhibit greater elongation and more active migration, which was probably due to the changes of adhesion strength and the actin stress fiber formation [42, 55, 56]. Moreover, it is possible that cells could feel the adhesion scale, and the clustered RGD peptides that had the larger scale in local nano-regions could provide more stable conjunctions for cells to stress their fibers on the cell–matrix interfaces [57, 58]. That could be the explanation for why C2C12 cells elongated longer in the highly RGD-clustered hydrogel than that in the homogenous case (Figure 6.8d). In addition, integrin clustering on cells usually plays the crucial role in regulating the rearrangement of its cytoskeletal structures, which will finally determine the activation and efficacy of its adhesion, migration, and motility [59–61]. RGD clustering leads to integrin aggregating on the cell–matrix interfaces, and regulates the mechanical forces between the substrate and cells during cell adhesion [55, 62]. It is important to control the spatial distribution and the concentration of the RGD peptides, which will alter the local mechanical properties of the substrate [38, 40–42]. Further, cells could feel and respond to such local mechanical properties, which will lead to the different cellular behaviors [63–65]. These are the empirical guidance for designing the ECM material.

6.4.3 Local Stiffness Variation Influenced by RGD Distributions in Hydrogels

The clustered RGD peptides decrease the amount of remaining sites on dextran polymer for crosslinking, which could influence the collective cellular behaviors in the 3D dextran hydrogel, including cell connection and aggregation. When the RGD peptides were clustered, the local stiffness of the dextran hydrogel may alter consequently (it can also be denoted as "stiffness clustering" in this study). This is because there are different amounts of available maleimide groups left on these two volume parts of polymers for the following crosslinking with CD-Link. Therefore, the dextran with clustered RGD peptides has a relatively weaker crosslinking strength than that without RGD peptides, which may cause the decrease of local mechanical strength within the hydrogel, which provided a relative softer ECM material for the adhered cells in it [66]. Additionally, previous efforts gave evidence that cells tended to extend their protrusion following the substrate long fibrils along their length, and such an effect would become more obvious when the fibrils were parallel to the direction of the protrusions [65]. When more and more cells adhered on an RGD clustered polymer, the polymer could not offer an adequate balance force for the cellular contraction force, due to the weak mechanical support by less stabilized crosslinking in the local positions with more RGD binding. One of the possible choices of cells is used to laterally (perpendicularly to the previous elongation direction) search other anchors for maintaining adhesion and spread according to the experimental observation. Therefore, C2C12 cells had more chance to exhibit more B-connections and less L-connections in higher RGD-clustered dextran hydrogel (Figures 6.10e and f). Additionally, such locally weak crosslinking may also enhance the aggregation of 3T3 cells. Due to the lowly crosslinking inside a local dextran substrate, the hydrogel could provide a looser, but still stable 3D substrate for 3T3 cells to aggregate into a bigger spherule more easily through degrading the linkers by excretive metalloproteinase (MMP). Hence, such a "stiffness-clustering" substrate may potentially function as the regulator of the size of 3T3 cell aggregation (Figures 6.10c and d).

In the homogenous case, because of the homogenous distribution of the RGD peptides, all the dextran polymer chains were crosslinked evenly. Therefore, the dextran hydrogel with homogenous RGD could provide relatively more mechanical stabilization spanning the multicellular scales for collective cell connections. That could possibly explain why C2C12 cells showed more L-connections and less B-connections in RGD-homogenous dextran hydrogels (Figures 6.10e and f). Additionally, such locally mechanical stable crosslinking may hinder the aggregation of 3T3 cells consequently. Due to the even crosslinking among the local dextran macromolecules, the hydrogel could provide a stable, but somewhat dense and tight 3D ECM for 3T3 cells, which may restrict them from aggregating into a bigger scale.

6.4.4 Significance of Cellular Behaviors Influenced by Averaged Concentration of RGD in Hydrogels

The 3D RGD-clustered dextran hydrogel usually showed above superiority remarkably when the averaged concentration of RGD was relatively lower (50 µM in this study). When the averaged concentration of RGD peptides increased greatly, the effects caused by clustered RGD would decline conspicuously. Too many anchors may hinder the active performance of cells in the hydrogel. We have tested the cases of 300-µM RGD peptides, and the results showed no superiority of C2C12 elongating in the RGD clustering cases after day 4 (Figure 6.11). Therefore, the advantages of RGD clustering hydrogel could be amplified by appropriately reducing the RGD concentration according to our results. The dextran hydrogel with clustered RGD peptides only requires a much smaller amount of RGD peptides for tuning the cellular spread, elongation, and aggregation. Further, the clustering of RGD may induce a more focused rearrangement of cytoskeletal fibers in cells by clustering bound integrins, and the hydrogel with clustered RGD could possibly provide more orientated footholds for cell adhesion and retraction. It can potentially become a competitive and economical method for stimulating the spread, elongation, aggregation, and migration of cells.

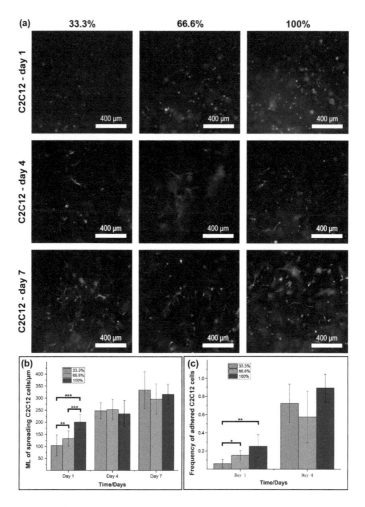

Figure 6.11 The impact of the composition configuration of 3D dextran hydrogels with clustered and homogenous RGD distribution on C2C12 cellular collective behaviors. The main parameters of dextran hydrogel were crosslinking strength = 2 mM and RGD = 300 μM. (a) F-actin of C2C12 cells were stained on days 1–7 for four different RGD clustering cases in hydrogels; cells were observed by an inverted microscope, and images were taken with a digital sCMOS camera; (b) The ML of spreading C2C12 cells was measured on days 1–7 for three different RGD clustering cases. The measurement was conducted by ImageJ software. (c) The frequency of adhered C2C12 cells was estimated on day 1 and day 4 for three different RGD clustering cases. The data was presented by mean ± SD; $*p < 0.05$, $**p < 0.03$, and $***p < 0.01$ versus the corresponding 3D homogenous samples.

6.4.5 Effect of Stiffness-Heterogeneity with Large Fluctuation and RGD Clustering Induced Stiffness-Heterogeneity with Small Variation

Further, recent work has also studied the impact of heterogeneity in hydrogel stiffness on cellular spreading, migration, or differentiation. For example, the cells in heterogeneously stiff substrates have been observed to elongate or migrate more toward the stiffer part of the substrate, which can be termed as durotaxis [67–69]. Durotaxis is one of the effective factors to guide cell directional motility [70]. For instance, the migratory response of 3T3 fibroblast cells was found consistent with the durotaxis prediction on the micro-patterned hydrogel [71] and previous work has proved the effect of durotaxis on the hydrogel with cell-scaled heterogeneous elasticity [72]. However, the built heterogeneous stiffness of matrix usually had remarkably larger dimensions than or close to that of individual cells (10 µm or larger) [73, 74]. In order to further explore the effect of nano- or molecular-scaled heterogeneous stiffness of hydrogel matrix on individual cellular behaviors, including morphology and phenotype, we also demonstrated another rational and easy-to-implement strategy [75] that can tune the stiffness fluctuation within a local area just covered by an individual cell. This method is different from the scheme that is stated and discussed in Section 6.4.3. RGD-clustering induced stiffness-heterogeneity usually involves smaller variation of stiffness because it is induced by the smaller variation of RGD concentration under the condition that the total common sites for binding RGD peptides and crosslinkers are kept as constant; while the method in literature [75] can offer a larger variation of stiffness within a single cell region by directly manipulating the amount proportion of crosslinkers in the two parts with equal quantity — lowly crosslinking part (L-part) and highly crosslinking part (H-part).

Although the different extents of stiffness-heterogeneity achieved by these two methods, each of them can find its own applications for controlling the behaviors of certain types of cells during their self-assembly in 3D hydrogels, since different types of cells may have distinct sensitivities and responses to a certain stiffness variation in nano-scaled regions [74]. For example, the myoblasts are sensitive to the stress, so they are susceptible to the RGD clustering induced stiffness-heterogeneity with small variation [76] and the direct crosslinker-apportion induced stiffness-heterogeneity with large fluctuation [75].

6.5 Summary

The 3D mal-dextran hydrogels with homogenous and clustered RGD compositions provided extracellular environments with different external stimuli for cells, and offered flexible and quantitative regulation for the collective cellular behaviors. Our results demonstrated that RGD-clustered mal-dextran hydrogel has the advantage of enhancing C2C12 cell elongation, the breadthwise-aggregated connection, and promoting the 3T3 cell aggregating degree relative to the sample with homogenous RGD. Further, the advantages of RGD clustering hydrogel could be amplified by appropriately reducing RGD concentration. The reason could probably be attributed to the influence of the RGD ligand distribution on the cell–matrix adhesion strength and the actin stress fiber formation, which then results in the more stable conjunctions for and greater polarized elongation of cells when the clustered RGD peptides are available. It also explained that C2C12 cells elongated longer in the highly RGD-clustered hydrogel than that in the homogenous case. The clustering rate of RGD peptides embedded in this hydrogel could function as one of the regulators of cellular growth behaviors such as cellular adhesion effect. Another rational and easy-to-implement strategy is direct crosslinker-apportion induced stiffness-heterogeneity with large fluctuation, which can tune the stiffness fluctuation within a local area covered by an individual cell. It is different from RGD-clustering induced stiffness-heterogeneity involving smaller variation, yet can also regulate the cellular behaviors during the self-organization of cells.

Acknowledgments

This work is supported by the National Natural Science Foundation of China (Grant Nos. 51875170 and 51505127) and Fundamental Research Funds for the Central Universities (Grant Nos. B200202225, 2018B22414 and 2015B04414) of China.

References

1. Mimeault, M.; Batra, S. K. (2006). Concise review: Recent advances on the significance of stem cells in tissue regeneration and cancer therapies, *Stem Cells,* **24**, pp. 2319–45.

2. Eiraku, M.; Takata, N.; Ishibashi, H.; Kawada, M.; Sakakura, E.; Okuda, S.; Sekiguchi, K.; Adachi, T.; Sasai, Y. (2011). Self-organizing optic-cup morphogenesis in three-dimensional culture, *Nature,* **472**, pp. 51–6.

3. Rizzoti, K.; Lovell-Badge, R. (2011). Regenerative medicine: Organ recital in a dish, *Nature,* **480**, pp. 44–6.

4. Bashur, C. A.; Venkataraman, L.; Ramamurthi, A. (2012). Tissue engineering and regenerative strategies to replicate biocomplexity of vascular elastic matrix assembly, *Tissue Engineering Part B: Reviews,* **18**, pp. 203–217.

5. Liu, J. S.; Gartner, Z. J. (2012). Directing the assembly of spatially organized multicomponent tissues from the bottom up, *Trends Cell Biol.,* **22**, pp. 683–91.

6. Suga, H.; Kadoshima, T.; Minaguchi, M.; Ohgushi, M.; Soen, M.; Nakano, T.; Takata, N.; Wataya, T.; Muguruma, K.; Miyoshi, H.; Yonemura, S.; Oiso, Y.; Sasai, Y. (2011). Self-formation of functional adenohypophysis in three-dimensional culture, *Nature,* **480**, pp. 57–62.

7. Takebe, T.; Sekine, K.; Enomura, M.; Koike, H.; Kimura, M.; Ogaeri, T.; Zhang, R. R.; Ueno, Y.; Zheng, Y. W.; Koike, N.; Aoyama, S.; Adachi, Y.; Taniguchi, H. (2013). Vascularized and functional human liver from an iPSC-derived organ bud transplant, *Nature,* **499**, pp. 481–484.

8. Athanasiou, K. A.; Eswaramoorthy, R.; Hadidi, P.; Hu, J. C. (2013). Self-organization and the self-assembling process in tissue engineering, *Annu. Rev. Biomed. Eng.,* **15**, pp. 115–136.

9. Tewary, M.; Shakiba, N.; Zandstra, P. W. (2018). Stem cell bioengineering: Building from stem cell biology, *Nature Reviews Genetics,* **19**, pp. 595–614.

10. Rossi, G.; Manfrin, A.; Lutolf, M. P. (2018). Progress and potential in organoid research, *Nature Reviews Genetics,* **19**, pp. 671–687.

11. Brusatin, G.; Panciera, T.; Gandin, A.; Citron, A.; Piccolo, S. (2018). Biomaterials and engineered microenvironments to control YAP/TAZ-dependent cell behaviour, *Nat. Mater.,* **17**, pp. 1063–1075.

12. Blache, U.; Vallmajo-Martin, Q.; Horton, E. R.; Guerrero, J.; Djonov, V.; Scherberich, A.; Erler, J. T.; Martin, I.; Snedeker, J. G.; Milleret, V. (2018). Notch-inducing hydrogels reveal a perivascular switch of mesenchymal stem cell fate, *EMBO reports,* **19**(8), p. e45964.

13. Kim, S.; Shah, S. B.; Graney, P. L.; Singh, A. (2019). Multiscale engineering of immune cells and lymphoid organs, *Nature Reviews Materials,* **4**(6), pp. 355–378.

14. Mirbagheri, M.; Adibnia, V.; Hughes, B. R.; Waldman, S. D.; Banquy, X.; Hwang, D. K. (2019). Advanced cell culture platforms: A growing quest for emulating natural tissues, *Materials Horizons,* **6**, pp. 45–71.

15. Elliott, N. T.; Yuan, F. (2011). A review of three-dimensional in vitro tissue models for drug discovery and transport studies, *J. Pharm. Sci.,* **100**, pp. 59–74.

16. Huang, C.; Shen, X.; Liu, X.; Chen, Z.; Shu, B.; Wan, L.; Liu, H.; He, J. (2019). Hybrid breath figure method: A new insight in Petri dishes for cell culture, *J. Colloid Interface Sci.,* **541**, pp. 114–122.

17. Duval, K.; Grover, H.; Han, L.-H.; Mou, Y.; Pegoraro, A. F.; Fredberg, J.; Chen, Z. (2017). Modeling physiological events in 2D vs 3D cell culture, *Physiology,* **32**, pp. 266–277.

18. Boehnke, N.; Cam, C.; Bat, E.; Segura, T.; Maynard, H. D. (2015). Imine hydrogels with tunable degradability for tissue engineering, *Biomacromolecules,* **16**, pp. 2101–2108.

19. Tibbitt, M. W.; Anseth, K. S. (2009). Hydrogels as extracellular matrix mimics for 3D cell culture, *Biotechnol. Bioeng.,* **103**, pp. 655–663.

20. Eglen, R. M.; Klein, J.-L. (2017). Three-dimensional cell culture: a rapidly emerging approach to cellular science and drug discovery. *SLAS Discovery,* **22**, pp. 453–455.

21. Caliari, S. R.; Burdick, J. A. (2016). A practical guide to hydrogels for cell culture, *Nat. Methods,* **13**, pp. 405.

22. Xu, T.; Molnar, P.; Gregory, C.; Das, M.; Boland, T.; Hickman, J. J. (2009). Electrophysiological characterization of embryonic hippocampal neurons cultured in a 3D collagen hydrogel, *Biomaterials,* **30**, pp. 4377–4383.

23. Fontana, F.; Raimondi, M.; Marzagalli, M.; Sommariva, M.; Limonta, P.; Gagliano, N. (2019). Epithelial-to-mesenchymal transition markers and CD44 Isoforms are differently expressed in 2D and 3D cell cultures of prostate cancer cells, *Cells,* **8**, pp. 1–17.

24. Fleischer, S.; Jahnke, H. G.; Fritsche, E.; Girard, M.; Robitzki, A. A. (2019). Comprehensive human stem cell differentiation in a 2D and 3D mode to cardiomyocytes for long-term cultivation and multiparametric monitoring on a multimodal microelectrode array setup, *Biosens Bioelectron,* **126**, pp. 624–631.

25. Chai, Q.; Jiao, Y.; Yu, X. (2017). Hydrogels for biomedical applications: Their characteristics and the mechanisms behind them, *Gels (Basel, Switzerland),* **3**, pp. 1–15.

26. Chirani, N.; Yahia, L. H.; Gritsch, L.; Motta, F. L.; Chirani, S.; Faré, S. (2016). History and applications of hydrogels, *J Biomedical Sci.,* **4**, pp. 13.

27. Tiwari, S.; Patil, R.; Bahadur, P. (2019). Polysaccharide based scaffolds for soft tissue engineering applications, *Polymers,* **11**, pp. 1.

28. Riahi, N.; Liberelle, B.; Henry, O.; De Crescenzo, G. (2017). Impact of RGD amount in dextran-based hydrogels for cell delivery, *Carbohydrate Polymers,* **161**, pp. 219–227.

29. Lei, Y.; Gojgini, S.; Lam, J.; Segura, T. (2011). The spreading, migration and proliferation of mouse mesenchymal stem cells cultured inside hyaluronic acid hydrogels, *Biomaterials,* **32**, pp. 39–47.

30. Adelöw, C.; Segura, T.; Hubbell, J. A.; Frey, P. (2008). The effect of enzymatically degradable poly (ethylene glycol) hydrogels on smooth muscle cell phenotype, *Biomaterials,* **29**, pp. 314–326.

31. Nuttelman, C. R.; Henry, S. M.; Anseth, K. S. (2002). Synthesis and characterization of photocrosslinkable, degradable poly (vinyl alcohol)-based tissue engineering scaffolds, *Biomaterials,* **23**, pp. 3617–3626.

32. Nikpour, P.; Salimi-Kenari, H.; Fahimipour, F.; Rabiee, S. M.; Imani, M.; Dashtimoghadam, E.; Tayebi, L. (2018). Dextran hydrogels incorporated with bioactive glass-ceramic: Nanocomposite scaffolds for bone tissue engineering, *Carbohydr. Polym.,* **190**, pp. 281–294.

33. Alsberg, E.; Anderson, K.; Albeiruti, A.; Franceschi, R.; Mooney, D. (2001). Cell-interactive alginate hydrogels for bone tissue engineering, *J. Dent. Res.,* **80**, pp. 2025–2029.

34. Lam, J.; Segura, T. (2013). The modulation of MSC integrin expression by RGD presentation, *Biomaterials,* **34**, pp. 3938–3947.

35. Cavo, M.; Fato, M.; Peñuela, L.; Beltrame, F.; Raiteri, R.; Scaglione, S. (2016). Microenvironment complexity and matrix stiffness regulate breast cancer cell activity in a 3D in vitro model, *Scientific Reports,* **6**, pp. 35367.

36. Discher, D. E.; Mooney, D. J.; Zandstra, P. W. (2009). Growth factors, matrices, and forces combine and control stem cells, *Science,* **324**, pp. 1673–1677.

37. Lutolf, M. P.; Gilbert, P. M.; Blau, H. M. (2009). Designing materials to direct stem-cell fate, *Nature,* **462**, pp. 433.

38. Hersel, U.; Dahmen, C.; Kessler, H. (2003). RGD modified polymers: Biomaterials for stimulated cell adhesion and beyond, *Biomaterials,* **24**, pp. 4385–4415.

39. Ouyang, L.; Dan, Y.; Shao, Z. W.; Yang, S. H.; Yang, C.; Liu, G. H.; Duan, D. Y. (2019). MMP-sensitive PEG hydrogel modified with RGD promotes

bFGF, VEGF and EPC-mediated angiogenesis, *Experimental and Therapeutic Medicine,* **18**, pp. 2933–2941.

40. Giancotti, F. G.; Ruoslahti, E. (1999). Integrin signaling, *Science,* **285**, pp. 1028–1033.

41. Comisar, W. A.; Kazmers, N. H.; Mooney, D. J.; Linderman, J. J. (2007). Engineering RGD nanopatterned hydrogels to control preosteoblast behavior: A combined computational and experimental approach, *Biomaterials,* **28**, pp. 4409–4417.

42. Comisar, W.; Mooney, D.; Linderman, J. (2011). Integrin organization: Linking adhesion ligand nanopatterns with altered cell responses, *J. Theor. Biol.,* **274**, pp. 120–130.

43. Van Tomme, S. R.; Hennink, W. E. (2007). Biodegradable dextran hydrogels for protein delivery applications, *Expert Review of Medical Devices,* **4**, pp. 147–164.

44. Neely, W. B. (1960). Dextran: Structure and synthesis, *Advances in Carbohydrate Chemistry,* **15**, pp. 341–69.

45. Zhang, Y.; Chu, C.-C. (2016). The effect of molecular weight of biodegradable hydrogel components on indomethacin release from dextran and poly(DL)lactic acid based hydrogels, *J. Bioact. Compatible Polym.,* **17**, pp. 65–85.

46. Giano, M. C.; Ibrahim, Z.; Medina, S. H.; Sarhane, K. A.; Christensen, J. M.; Yamada, Y.; Brandacher, G.; Schneider, J. P. (2014). Injectable bioadhesive hydrogels with innate antibacterial properties, *Nature Communications,* **5**, pp. 4095.

47. White, J. A.; Deen, W. M. (2002). Agarose-dextran gels as synthetic analogs of glomerular basement membrane: Water permeability, *Biophys. J.,* **82**, pp. 2081–2089.

48. Liu, Z. Q.; Wei, Z.; Zhu, X. L.; Huang, G. Y.; Xu, F.; Yang, J. H.; Osada, Y.; Zrinyi, M.; Li, J. H.; Chen, Y. M. (2015). Dextran-based hydrogel formed by thiol-Michael addition reaction for 3D cell encapsulation, *Colloids and Surfaces B-Biointerfaces,* **128**, pp. 140–148.

49. van Dijk-Wolthuis, W. N. E.; Franssen, O.; Talsma, H.; van Steenbergen, M. J.; Kettenes-van den Bosch, J. J.; Hennink, W. E. (1995). Synthesis, characterization, and polymerization of glycidyl methacrylate derivatized dextran, *Macromolecules,* **28**, pp. 6317–6322.

50. Hiemstra, C.; van der Aa, L. J.; Zhong, Z.; Dijkstra, P. J.; Feijen, J. (2007). Novel in situ forming, degradable dextran hydrogels by Michael addition chemistry: Synthesis, rheology, and degradation, *Macromolecules,* **40**, pp. 1165–1173.

51. Hiemstra, C.; van der Aa, L. J.; Zhong, Z.; Dijkstra, P. J.; Feijen, J. (2007). Rapidly in situ-forming degradable hydrogels from dextran thiols through michael addition, *Biomacromolecules*, **8**, pp. 1548–1556.

52. vanDijkWolthuis, W. N. E.; Hoogeboom, J. A. M.; vanSteenbergen, M. J.; Tsang, S. K. Y.; Hennink, W. E. (1997). Degradation and release behavior of dextran-based hydrogels, *Macromolecules*, **30**, pp. 4639–4645.

53. The buffer system, user guide, support, https://www.cellendes.com/.

54. Orsi, G.; Fagnano, M.; De Maria, C.; Montemurro, F.; Vozzi, G. (2017). A new 3D concentration gradient maker and its application in building hydrogels with a 3D stiffness gradient, *J. Tissue Eng. Regen. Med.*, **11**, pp. 256–264.

55. Ladoux, B.; Mège, R.-M. (2017). Mechanobiology of collective cell behaviours, *Nature Reviews Molecular Cell Biology*, **18**, pp. 743.

56. Maheshwari, G.; Brown, G.; Lauffenburger, D. A.; Wells, A.; Griffith, L. G. (2000). Cell adhesion and motility depend on nanoscale RGD clustering, *J. Cell Sci.*, **113**, pp. 1677–1686.

57. Svendsen, K. H.; Thomson, G. (1984). A new clamping and stretching procedure for determination of collagen fiber stiffness and strength relations upon maturation, *J. Biomech.*, **17**, pp. 225–229.

58. Doyle, A. D.; Carvajal, N.; Jin, A.; Matsumoto, K.; Yamada, K. M. (2015). Local 3D matrix microenvironment regulates cell migration through spatiotemporal dynamics of contractility-dependent adhesions, *Nature Communications*, **6**, pp. 8720.

59. Kanchanawong, P.; Shtengel, G.; Pasapera, A. M.; Ramko, E. B.; Davidson, M. W.; Hess, H. F.; Waterman, C. M. (2010). Nanoscale architecture of integrin-based cell adhesions, *Nature*, **468**, pp. 580.

60. Behrndt, M.; Salbreux, G.; Campinho, P.; Hauschild, R.; Oswald, F.; Roensch, J.; Grill, S. W.; Heisenberg, C.-P. (2012). Forces driving epithelial spreading in zebrafish gastrulation, *Science*, **338**, pp. 257–260.

61. Abreu-Blanco, M. T.; Verboon, J. M.; Liu, R.; Watts, J. J.; Parkhurst, S. M. (2012). Drosophila embryos close epithelial wounds using a combination of cellular protrusions and an actomyosin purse string, *J. Cell Sci.*, **125**, pp. 5984–5997.

62. Schwarz, U. S.; Gardel, M. L. (2012). United we stand –integrating the actin cytoskeleton and cell–matrix adhesions in cellular mechanotransduction, *J. Cell Sci.*, **125**, pp. 3051–3060.

63. Ridley, A. J.; Schwartz, M. A.; Burridge, K.; Firtel, R. A.; Ginsberg, M. H.; Borisy, G.; Parsons, J. T.; Horwitz, A. R. (2003). Cell migration: Integrating signals from front to back, *Science*, **302**, pp. 1704–1709.

64. Lauffenburger, D. A.; Horwitz, A. F. (1996). Cell migration: A physically integrated molecular process, *Cell,* **84**, pp. 359–369.

65. Balaban, N. Q.; Schwarz, U. S.; Riveline, D.; Goichberg, P.; Tzur, G.; Sabanay, I.; Mahalu, D.; Safran, S.; Bershadsky, A.; Addadi, L. (2001). Force and focal adhesion assembly: A close relationship studied using elastic micropatterned substrates, *Nat. Cell Biol.,* **3**, pp. 466.

66. Pedersen, J. A.; Swartz, M. A. (2005). Mechanobiology in the third dimension, *Annals of Biomedical Engineering,* **33**, pp. 1469–1490.

67. Lachowski, D.; Cortes, E.; Pink, D.; Chronopoulos, A.; Karim, S. A.; Morton, J. P.; Armando, E. (2017). Substrate rigidity controls activation and durotaxis in pancreatic stellate cells, *Scientific Reports,* **7**, pp. 2506.

68. Hartman, C. D.; Isenberg, B. C.; Chua, S. G.; Wong, J. Y. (2016). Vascular smooth muscle cell durotaxis depends on extracellular matrix composition, *Proceedings of the National Academy of Sciences,* **113**, pp. 11190–11195.

69. Sunyer, R.; Conte, V.; Escribano, J.; Elosegui-Artola, A.; Labernadie, A.; Valon, L.; Navajas, D.; García-Aznar, J. M.; Muñoz, J. J.; Roca-Cusachs, P. (2016). Collective cell durotaxis emerges from long-range intercellular force transmission, *Science,* **353**, pp. 1157–1161.

70. Sunyer, R.; Trepat, X. (2020). Durotaxis, *Current Biology,* **30**, pp. R383–R387.

71. Kuboki, T.; Chen, W.; Kidoaki, S. (2014). Time-dependent migratory behaviors in the long-term studies of fibroblast durotaxis on a hydrogel substrate fabricated with a soft band, *Langmuir the Acs Journal of Surfaces & Colloids,* **30**, pp. 6187.

72. Ebata, H.; Moriyama, K.; Kuboki, T.; Kidoaki, S. (2020). General cellular durotaxis induced with cell-scale heterogeneity of matrix-elasticity, *Biomaterials,* **230**, pp. 119647.

73. Li, S.; Bai, H.; Chen, X.; Gong, S.; Xiao, J.; Li, D.; Li, L.; Jiang, Y.; Li, T.; Qin, X.; Yang, H.; Wu, C.; You, F.; Liu, Y. (2020). Soft substrate promotes osteosarcoma cell self-renewal, differentiation, and drug resistance through miR-29b and its target protein spin 1, *ACS Biomaterials Science & Engineering,* **6**, pp. 5588–5598.

74. Deng, J.; Zhao, C.; Spatz, J. P.; Wei, Q. (2017). Nanopatterned adhesive, stretchable hydrogel to control ligand spacing and regulate cell spreading and migration, *ACS nano,* **11**, pp. 8282–8291.

75. Wang, Z.; Zhu, X.; Yin, X. (2020). Quantitatively designed cross-linker-clustered maleimide–dextran hydrogels for rationally regulating the behaviors of cells in a 3D matrix, *ACS Applied Bio Materials,* **3**, pp. 5759–5774.

76. Wang, Z.; Zhu, X.; Zhang, R. (2019). Characterization and analysis of collective cellular behaviors in 3D dextran hydrogels with homogenous and clustered RGD compositions, *Materials,* **12**, 3391: 1–23.

Index